# Precise Energy
## A Missing View on Batteries

# Precise Energy
## A Missing View on Batteries

Editors

### Kai Peter Birke
*University of Stuttgart & Fraunhofer Institute for Manufacturing Engineering and Automation, Germany*

### Sabri Baazouzi
*Fraunhofer Institute for Manufacturing Engineering and Automation, Germany*

### Julian Joël Grimm
*Fraunhofer Institute for Manufacturing Engineering and Automation, Germany*

*Published by*

World Scientific Publishing Co. Pte. Ltd.

5 Toh Tuck Link, Singapore 596224

*USA office:* 27 Warren Street, Suite 401-402, Hackensack, NJ 07601

*UK office:* 57 Shelton Street, Covent Garden, London WC2H 9HE

Library of Congress Control Number: 2024038425

**British Library Cataloguing-in-Publication Data**
A catalogue record for this book is available from the British Library.

**PRECISE ENERGY**
**A Missing View on Batteries**

Copyright © 2025 by World Scientific Publishing Co. Pte. Ltd.

*All rights reserved. This book, or parts thereof, may not be reproduced in any form or by any means, electronic or mechanical, including photocopying, recording or any information storage and retrieval system now known or to be invented, without written permission from the publisher.*

For photocopying of material in this volume, please pay a copying fee through the Copyright Clearance Center, Inc., 222 Rosewood Drive, Danvers, MA 01923, USA. In this case permission to photocopy is not required from the publisher.

ISBN 9789811282041 (hardcover)
ISBN 9789811282058 (ebook for institutions)
ISBN 9789811282065 (ebook for individuals)

For any available supplementary material, please visit
https://www.worldscientific.com/worldscibooks/10.1142/13562#t=suppl

Desk Editors: Soundararajan Raghuraman/Amanda Yun

Typeset by Stallion Press
Email: enquiries@stallionpress.com

# About the Editors

**Kai Peter Birke** is a physicist, materials scientist, and a full professor at the University of Stuttgart, Germany, covering the field of electrical energy storage systems, including new energy storage cell materials and technologies, advanced Li-ion batteries, and Power-to-X. He obtained his PhD in materials science (ion conducting ceramics) from the University of Kiel, Germany, in 1998. In 1999, he joined the Fraunhofer Institute for Silicon Technology, Itzehoe, Germany, to work on the development of proprietary Li-ion laminated hybrid solid cells with novel functional ceramic separators. He also co-founded two spin-offs to put this technology into production. After being involved in the development and production of pouch-type laminated Li-ion cells (PoLiFlex) in leading positions at VARTA AG for five years (2000–2005), Prof. Birke joined Continental AG, Business Unit Hybrid Electric Vehicle, in Berlin in 2005, as a senior expert and project leader in energy storage systems. He subsequently became a senior technical expert in battery technology and team leader in cell technology, and in 2010, he was appointed Head of Battery Modules and Electromechanics. In 2013, he joined the Joint Venture (JV) SK-Continental E-motion as Head of Advanced Development. He was one of the key pioneers for this JV. In 2015, Prof. Birke became a full professor at the University of Stuttgart, Institute for Photovoltaics (IPV), as head of electrical energy storage systems. Additionally, he also leads the Center for Digitalized Battery Cell Manufacturing at Fraunhofer Institute for Manufacturing Engineering and Automation (Fraunhofer IPA), with an extended focus on battery cell production and digitalization.

**Sabri Baazouzi** is a specialist in mechanical and manufacturing engineering. In 2020, he joined the Fraunhofer Institute for Manufacturing Engineering and Automation, working at the Center for Digitalized Battery Cell Manufacturing. Since then, he has successfully developed and coordinated multiple research projects in collaboration with industry and academic partners. His research focuses on battery cell production and circular economy strategies for battery systems.

Mr. Baazouzi served as the project manager for the DeMoBat project, which concentrated on developing industrial automated disassembly solutions for battery systems. Additionally, he co-developed a fully automated and digitalized process chain to enable the flexible manufacturing of cylindrical battery cells with customizable formats and designs.

In July 2024, Mr. Baazouzi completed his PhD at the University of Stuttgart, where his research centered on design-flexible production methods within the battery ecosystem. Since March 2024, he has been working as a consultant at Mercedes-Benz, supporting the establishment of gigafactories for battery cell production across Europe.

**Julian Joël Grimm** is an industrial engineer with a specialization in renewable energies and energy efficiency and manufacturing engineering. Currently, he is working as a research team manager for battery technologies and project leader at the Fraunhofer Institute for Manufacturing Engineering and Automation (Fraunhofer IPA) in Stuttgart, Germany. He is also deputy head of the Center for Digitalized Battery Cell Manufacturing. The main topics of the working group are manufacturing technologies for battery cells and developing digital adding-value services for battery cell manufacturing, future battery materials and designs and their implications on manufacturing technologies, as well as developing a circular economy for batteries with end-of-life strategies and solutions for direct recycling for manufacturing waste. At the Center for Digitalized Battery Cell Manufacturing, a pilot line for producing design- and format-flexible cylindrical cells has been implemented and is continuously developed. Between 2017 and 2019, Julian Grimm worked as a consultant for manufacturing engineering and logistics. Prior to that, he studied industrial engineering and completed his master's degree (MSc) at the University of Kassel (Universität Kassel), Germany.

© 2025 World Scientific Publishing Company
https://doi.org/10.9789811282058_fmatter

# About the Contributors

**Yucheng Luo** has been working as a research associate, specializing in battery modeling algorithms and digitalized battery production, at the Fraunhofer Institute for Manufacturing Engineering and Automation IPA in Stuttgart, Germany, since 2023. Prior to this, he was employed by BatterieIngenieure GmbH from 2021 to 2023, where he worked in a team focused on battery testing, modeling, and diagnostics. Before that, he completed his master of science (MSc) in sustainable energy supply at RWTH Aachen University in Germany.

**Daniel Steffen Reichert** is part of the Center for Digitalized Battery Cell Manufacturing at the Fraunhofer Institute for Manufacturing Engineering and Automation IPA in Stuttgart, Germany. He is actively engaged in various aspects of battery cell production research, focusing on the upscaling and industrialization of sodium-ion battery technology. Previously, Reichert studied business chemistry at the University of Ulm, Germany.

**Paul Rößner** is a chemist specializing in $CO_2$ electrolysis and plasma-based $CO_2$ conversion. After earning a PhD in chemical technology, Dr. Rößner joined the Institute for Photovoltaics at the University of Stuttgart, Germany, to lead the Power-to-X Group within the chair of energy storage systems. As the scientific coordinator of the CHEMampere initiative, he leads scientific efforts to electrify the chemical industry. His research focuses on plasma catalysis and plasma gas conversion technologies to transform $CO_2$, $N_2$, and water into valuable resources using renewable electricity. Dr. Rößner contributes practical approaches and innovative solutions to the "Power-to-X" field.

**Kathrin Schad** studied process engineering, specializing in chemical process engineering and energy process engineering, at the University of Stuttgart, Germany, where she completed her MSc in 2020 with a thesis on the optimization of a fast-charging profile of lithium-ion batteries. Since 2021, she has been working at the Fraunhofer Institute for Manufacturing Engineering and Automation, Germany, as a research associate and PhD student. Her work focuses on the development of next-generation batteries, such as all-solid-state batteries and alloy-based anode technologies, and their scaling to different cell formats.

**Elisa Thauer** is a research associate specializing in battery technology at the Fraunhofer Institute for Manufacturing Engineering and Automation, Germany. She studied physics at the University of Heidelberg, Germany, where she completed her BSc in 2014 with a thesis on the characterization of new electrode materials for Li-ion batteries. In 2017, she obtained her MSc degree, focusing on the influence of electrochemical de/lithiation on the magnetic properties and structure of materials. She received her Dr. rer. nat. in physics from the same university in 2021. Her research entailed the targeted optimization of electrode materials for Li- and

Na-ion batteries, investigating the influence of morphology and functionalization on the electrochemical properties.

**Johannes Wanner** is part of the Center for Battery Cell Manufacturing at the Fraunhofer Institute for Manufacturing Engineering and Automation IPA in Stuttgart, Germany, where he is engaged in battery cell testing, electrolyte filling, formation, and related cell quality issues in battery cell manufacturing. Previously, Wanner studied electrical engineering and information technology as well as electromobility at the University of Stuttgart, Germany.

© 2025 World Scientific Publishing Company
https://doi.org/10.9789811282058_fmatter

# Contents

| | |
|---|---|
| *About the Editors* | v |
| *About the Contributors* | vii |
| *Acknowledgments* | xiii |
| *Introduction – Energy Reimagined: Precise Discussions on Batteries are Shaping the Future*<br>*Julian Joël Grimm and Kai Peter Birke* | xv |

| | | |
|---|---|---|
| Chapter 1 | The Periodic System of Elements as the Limiting Factor for Galvanic Cell Design<br>*Elisa Thauer and Kai Peter Birke* | 1 |
| Chapter 2 | Energy Density vs. Power Density: Implications for Electrolyte Quantity and Filling Procedure<br>*Johannes Wanner and Kai Peter Birke* | 21 |
| Chapter 3 | The Best Battery Cell Candidates for Highest Energy Density<br>*Kathrin Schad and Kai Peter Birke* | 41 |
| Chapter 4 | Energy Density versus Power Density, Lifespan, Safety, and Costs<br>*Sabri Baazouzi and Kai Peter Birke* | 93 |
| Chapter 5 | The Role of Raw Materials in Enhancing or Limiting Energy Density<br>*Daniel Steffen Reichert and Kai Peter Birke* | 139 |

xii *Contents*

| | | |
|---|---|---|
| Chapter 6 | Remaining Energy Densities at the Battery Systems Level<br>*Julian Joël Grimm and Kai Peter Birke* | 187 |
| Chapter 7 | Achievable Progress in Battery Energy Densities<br>*Yucheng Luo and Kai Peter Birke* | 201 |
| Chapter 8 | Power-to-X Technology for Decarbonizing Non-Land Transportation, Heating, and Chemical Industry<br>*Paul Rößner and Kai Peter Birke* | 221 |
| *Index* | | 255 |

© 2025 World Scientific Publishing Company
https://doi.org/10.9789811282058_fmatter

# Acknowledgments

The editors would like to thank everybody who supported the preparation of this book.

Foremost, our thanks are due to the chapter authors. Their dedication not only to developing the content but also to engaging in numerous stimulating debates and sharing innovative ideas has been invaluable in shaping this work.

On behalf of all the authors, we also express our sincere appreciation to Eugenia Komnik for her meticulous proofreading of the chapters and her insightful suggestions.

© 2025 World Scientific Publishing Company
https://doi.org/10.9789811282058_fmatter

# Introduction – Energy Reimagined: Precise Discussions on Batteries are Shaping the Future

### Julian Joël Grimm[*] and Kai Peter Birke[†]

*Fraunhofer IPA, Nobelstrasse 12, Stuttgart, Germany*

*[*]julian.grimm@ipa.fraunhofer.de*

*[†]kai.peter.birke@ipa.fraunhofer.de*

In a world increasingly dependent on electric mobility and renewable energy, batteries stand at the forefront of technological progress. *Precise Energy – A Missing View on Batteries* highlights the crucial aspects often overlooked in the development and use of batteries. Specifically, batteries cannot exceed certain limits of energy density. Additionally, for all-solid-state batteries, a conflict may arise between energy density and power density. Traditionally, the discussion focuses on energy and power density. However, we must take an important step further by emphasizing circular value creation. This includes strategies such as reusing, repairing, repurposing, remanufacturing, and recycling. Therefore, decisions regarding future battery technologies should consider the crucial issue of sustainable and available raw materials.

Developing batteries that are not only powerful but also sustainable and precisely tailored to specific applications requires a shift in thinking. Future battery cells must be designed to meet the requirements of various processes within a circular value creation framework, as illustrated in

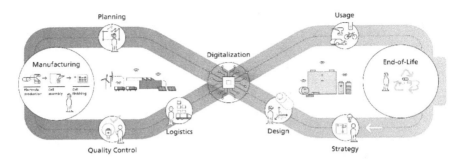

**Figure 1.** Circular value creation in the battery ecosystem.
*Source*: Copyright Fraunhofer IPA.

Figure 1. This approach ensures that environmental impacts are reduced and economic efficiency is increased, covering the entire lifecycle from development and manufacturing to usage, reuse, and recycling.

Another crucial factor is the quality of individual battery cells, especially when envisioning the so-called GWh factories. Each battery comprises numerous individual cells that must be consistently of high quality to achieve both performance and sustainability goals. For instance, if 10% of one billion manufactured battery cells are defective, this results in a waste of 100 million cells. To put this into perspective, one billion battery cells are required to produce one million battery electric vehicles, assuming an average of 1000 battery cells per vehicle. With the global annual production of nearly 100 million vehicles potentially being electrified, the necessity to conserve both natural and financial resources becomes even more pressing.

Digitalization is pivotal in minimizing these enormous waste quantities. By leveraging precise data analysis and integration such as machine learning, predictive analytics, and neural networks, intelligent systems can be developed to intervene early in the production process. These systems can also network data from stakeholders within the circular value creation framework to enhance overall value. In addition to digitalization, which addresses avoidable waste, direct recycling of production waste on the production line is crucial for managing unavoidable waste. This approach not only helps in reducing waste but also provides a competitive edge by optimizing resource utilization and improving overall sustainability.

Raw materials are critically important in this context, influencing not only the prices and sustainability of battery manufacturing but also the

*Introduction – Energy Reimagined*  xvii

recycling processes. In the years following the expansion of gigafactories, it is expected that the ratio of waste generated will shift from production waste to end-of-life batteries. After batteries are reused through circular strategies such as reusing, repairing, repurposing, or remanufacturing, sustainable and efficient recycling processes are essential for obtaining cost-effective and high-quality recycled materials. Material systems for lithium-ion battery cells, such as those based on lithium iron phosphate cathodes or sodium-ion battery cells, hold promise for manufacturing more sustainable and cost-effective batteries. However, these more affordable materials pose a challenge in recycling because their lower market value results in reduced revenue for recycling companies compared to more valuable material systems such as nickel manganese cobalt oxide. Despite this challenge, these battery technologies are essential for maintaining sustainability and cost-effectiveness and securing production capabilities in Germany and Europe.

"Precise Energy" is more than just a title. It embodies a concept with wide-reaching impacts. It involves designing batteries with such precision that they are tailored for their intended applications and fit seamlessly into circular value creation. It also means developing precise manufacturing processes to reduce waste, using energy efficiently, and optimizing resource utilization to foster a sustainable future. This book provides a comprehensive perspective on the challenges and opportunities within the battery industry, illustrating how we can secure a sustainable and prosperous future through careful and intentional actions. The combination of precise energy, innovative cell concepts, and sustainable circular value creation processes lays the groundwork for a new era in battery technology.

To transition into this new era effectively, it is essential to initiate a discussion about what batteries, as we currently understand them, can and cannot achieve. There will be no miraculous solutions – only steady, ongoing evolution. We also need a consistent view of what lies beyond batteries within a circular energy supply chain.

© 2025 World Scientific Publishing Company
https://doi.org/10.9789811282058_0001

# Chapter 1

# The Periodic System of Elements as the Limiting Factor for Galvanic Cell Design

**Elisa Thauer\* and Kai Peter Birke†**

*Fraunhofer IPA, Nobelstrasse 12, Stuttgart, Germany*

*\*elisa.thauer@ipa.fraunhofer.de*

*†kai.peter.birke@ipa.fraunhofer.de*

In this chapter, we discuss a fundamental issue for batteries, which is the given limitation in energy density. Galvanic cells, the electrochemical workhorses, which convert chemical energy into electrical energy via spontaneous redox (reduction-oxidation) reactions, depend strongly on the materials forming their electrodes and electrolytes. The periodic system of elements offers a comprehensive catalogue of materials from which scientists and engineers can draw. However, the periodic system also presents limits and challenges that must be considered in the development of high-performance, safe, and durable batteries. No additional element than helium can be invented between hydrogen and lithium. We show that these limitations are based on inherent properties of the periodic system of elements. Among those are electronegativity and electropositivity — properties represented by elements such as lithium (one of the most electropositive ones) and fluorine (the most electronegative one). No compounds can surpass the specific properties of these elements regarding electron structure and thus electronegativity and electropositivity. In other words, if we consider battery cells, they cannot exceed specific voltages since their storage principle is based on chemical potential (redox reactions) rather than an electric field, as is the case with capacitors. Another issue is to make lightweight elements, respectively, ions, such as $H^+$, $Li^+$, and $Na^+$ ions, ready for highly reversible rechargeable battery cells. This requires many more heavy elements, resulting in the fact that, for example, at the end, only about 3 wt.% metallic lithium is present in Li-ion cells. These critical and non-bridgeable facts, probably not in people's minds due to the visibility of electromobility, deserve thorough discussion in the opening chapter of this book about precise energy.

## 1.1 Electronegativity and Electropositivity as Limiting Factors

As previously mentioned, one limitation imposed by the periodic systems of elements (Figure 1.1) is a consequence of electronegativity and electropositivity, which describe the tendency of atoms to attract or donate electrons in a bond, respectively. This limits the voltage of an electrochemically driven cell to about 5.9 V, as the following calculation shows.

Let us look at opposite extremes in terms of electronegativity and electropositivity, namely lithium[1] and fluorine, and calculate the theoretical cell voltage of a battery that combines these two half-cells [1]:

$$\text{lithium (Li)} \quad Li \rightarrow Li^+ + e^-, \quad \text{potential: } -3.04 \text{ V;}$$

$$\text{fluorine (F)} \quad F_2 + e^- \rightarrow 2F^-, \quad \text{potential: } +2.87 \text{ V.}$$

The theoretical total voltage of a cell that combines these two half-cells would be the difference between the two potentials:

$$\Delta E = E_{\text{Cathode}} - E_{\text{Anode}} = (+2.87 \text{ V}) - (-3.04 \text{ V}) = 5.91 \text{ V}$$

It is not possible to find materials, either elements or compounds, that do not obey this rule. This is the reason why the story of Li-ion cells ends with the so-called "5 V materials."

---

[1]The electrode potential of francium, the most electropositive element, is even lower at $-3.05$ V, but as it is extremely rare and highly radioactive, it is not commonly used and thus not considered here.

The Periodic System of Elements as the Limiting Factor for Galvanic Cell Design  3

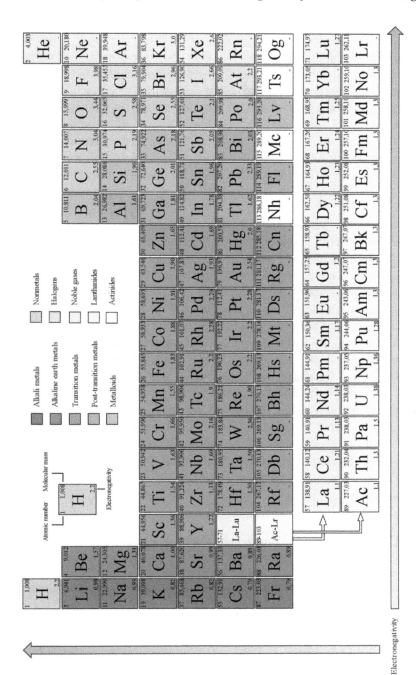

**Figure 1.1.** Periodic system of elements.

## 1.2 The Atomic Shelf as Limiting Factor

Comparable to a shelf where people reversibly place items for storage, modern rechargeable batteries allow ions to occupy an atomic "shelf" and store them with extremely high reversibility, similar to water in a sponge.

In nickel–metal hydride cells, protons are placed on this atomic shelf, while in modern Li-ion and Na-ion cells, lithium and sodium ions take their place. Other alkaline ions are not feasible since they stretch the host lattice too much due to their increased size [2]. Other small ions such as those of magnesium, calcium, aluminum, or silicon are multivalent and thus struggle severely with insufficient mobility at room temperature in host matrices [3].

However, the perfect shelf is a prerequisite for long-term, lasting rechargeable batteries. Leaving this basic principle of so-called "rocking-chair"-based cells, such as Li-ion batteries (LiBs), rechargeability suffers. As a result, such batteries are not competitive enough if the focus is on large energy throughputs over their lifetime.

This important atomic shelf is examined in detail in the following. Just to bring in mind the importance of the atomic shelf for rechargeability while considering its influence on the energy density, ask yourself the question: Has the hydrogen shuttle battery, which is based on the same principle as the Li- and Na-ion batteries and utilizes the lightweight element H for potentially higher capacity, already been invented? Yes, it is called nickel–metal hydride, the predecessor of the LiB. The atomic shelf matters so much that, in combination with a voltage of about only 1.2 V for such cells [4], there is insufficient competitiveness with Li-ion or even Na-ion. Let us therefore look at the atomic shelf in detail. Why is it necessary and how does it affect battery performance in terms of reversibility and capacity?

### 1.2.1 *Characteristics of batteries: Specific capacity and energy density*

Before diving into this topic, we first introduce the relevant performance characteristics of batteries, such as specific capacity and energy density. The specific capacity of an electrode is a key parameter in determining the performance of batteries. It is a measure of the amount of charge that can be stored per unit mass of an active material (AM) of the electrode. It is typically given in units of ampere-hours per gram (mAh/g). The specific capacity depends on the number of electrons transferred during the electrochemical redox reaction ($z$), the Faraday constant ($F$), and the molar mass of the active material ($M_{AM}$), which can be easily derived from the periodic system (Figure 1.1):

$$C^M = \frac{(z \times F)}{M_{AM}} \quad \left[ C = \frac{As}{g} \right], \tag{1.1}$$

where
$z$ is the number of electrons transferred,
$F$ is the Faraday constant (equal to 96.485 C/mol),
$M_{AM}$ is the molar mass of the active material in g/mol.

To calculate the specific capacity $C^M$ in the appropriate unit [mAh/g], equation (1.1) must be divided by a factor of 3.6 since 1 Coulomb is equal to 1 Ampere multiplied by 1 second.

Taking graphite as an example, which has indisputably emerged as the primary anode material in LIBs since the inception of commercial lithium-ion battery utilization in the 1990s and dominates the market with a market share of 98% [5]. Graphite can store Li ions in the maximum stoichiometry of $LiC_6$. This means that one lithium ion is intercalated for every six carbon atoms. The valence $z$ is therefore 1/6. The specific capacity of graphite can be calculated as follows:

$$C^M_{Graphite} = \frac{(1/6 \times 9.6485)}{12.001 \times 3,6} \approx 372 \frac{mAh}{g}.$$

The use of metallic lithium as an anode material in LIBs is a widely debated subject in battery research and development [6, 7]. This is because, theoretically, lithium has one of the highest specific capacities among all available materials. Its valence $z$ is 1, as one lithium atom emits one electron:

$$C^M_{Lithium} = \frac{(1 \times 96.485)}{6,941 \times 3,6} \approx 3.861 \frac{mAh}{g}.$$

The capacity of a whole electrochemical cell, which is composed of an anode and a cathode, represents the total amount of electric charge that the cell can store or deliver, and it is determined by the inverse sum of the capacities of the anode and cathode:

$$\frac{1}{C_{\text{Cell}}} = \frac{1}{C_{\text{Cathode}}} + \frac{1}{C_{\text{Anode}}}.$$

(1.2)

This equation is similar to that for the series connection of capacitors in electrical circuits. It reflects the fact that the overall capacity of the cell is influenced by the individual capacities of its electrodes and is essentially limited by the electrode with the lowest capacity, as the total capacity cannot exceed the capacity of the weakest electrode.

In addition to the specific capacity, the energy density is an important parameter for describing battery performance. The energy density expresses the amount of energy that can be stored in a cell. When discussing the capabilities of energy storage systems, a distinction is typically made between volumetric and gravimetric energy densities.

In specific cases, it can be extremely instructive to look at both kinds of energy density. As a rule of thumb, for many types of batteries, specifically Li-ion, the volumetric energy density is twice the gravimetric energy density. This is a tremendously important insight for electromobility and the reason that battery-powered electric vehicles are currently an available option. This does not apply to lithium-sulfur batteries. Here, volumetric and gravimetric energy densities are nearly equal. Therefore, lithium-sulfur is a limited technology if volumetric energy matters but remains promising technology if gravimetric energy density prevails, particularly in aerospace applications.

If new materials and approaches should be valued, it is extremely important to comment on both kinds of energy densities. Sometimes, the volumetric energy density may promise very attractive values, but the gravimetric energy density is less competitive. This may hold true nowadays for many so-called solid-state batteries.

Usually, the gravimetric energy density $E^{\text{M}}$ is decisive and is a good approach to identifying different battery technologies at a glance. It describes the amount of energy that can be stored per unit mass and is given in watt-hours per gram (Wh/g). It can be calculated from the specific capacity $C_{\text{Cell}}^{\text{M}}$ and the operating voltage $E_{\text{Cell}}$ of the cell as follows:

$$E^{\text{M}} = C_{\text{Cell}}^{\text{M}} \times E_{\text{Cell}}.$$

(1.3)

Modern batteries are currently approaching 200 Wh/kg on a system level [8]. Comparing this with the gravimetric energy densities of a large variety of fuels, it seems that there is a huge gap here. This is the reason why battery-powered electromobility has evolved from a "no-go" technology to one that is considered borderline but not a breakthrough technology. Actually, the energy density also hampers the positive cost development of battery-powered vehicles. If the energy density, based on similar materials, were to double, it would halve the costs in €/kWh and make electromobility much more attractive. Also, the weight remains an issue, e.g., for tire technology and the achievable kilometers per charge. Nevertheless, though the energy density of batteries is comparatively still quite low, battery-powered vehicles have become an option. The potential driver in the direction of partial electromobility will certainly be the cost per kWh. If these ultimately remain at the level of gasoline or diesel, which is definitely the case at 50–70 €ct/kWh, nothing will really move toward a breakthrough for electromobility. However, if about 20–25 €ct/kWh become feasible (as actually politically supported in Germany for heat pumps [9]), this feature could become a real game changer.

### 1.2.2 Energy yield of elements: Internal combustion engine vs. oxygen battery

In this section, the energy yield of elements is examined when used as (hypothetical) fuels in an internal combustion engine and as cathode material in an oxygen battery. What are the differences?

An internal combustion engine burns fuel to generate mechanical energy. It draws in air from the environment, which is necessary for combustion, and ignites the fuel–air mixture, converting chemical energy into mechanical energy through an explosion. The resulting exhaust gases are then released into the atmosphere. Figure 1.2 illustrates the reactions that take place when different (hypothetical) fuels are used. The combustion of hexane, which is used here as a model compound for petrol and diesel, produces carbon dioxide ($CO_2$) and water ($H_2O$). In contrast, when hydrogen ($H_2$) is used as a fuel, only water vapor is produced as exhaust gas.

**Excursus – Impact of the exhaust gases on the environment:** Burning fossil fuels, such as petrol or diesel, produces carbon dioxide, water vapor, and other exhaust gases. The impact of $CO_2$ on the environment is well

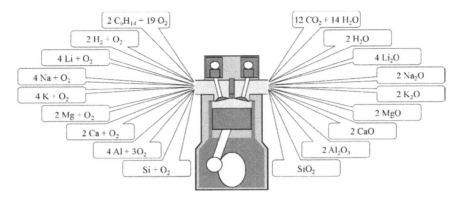

**Figure 1.2.** Sketch of an internal combustion engine running on various (hypothetical) fuels with the corresponding combustion reaction.

known; however, water as a by-product of combustion must also be viewed critically. It is not the main problem per se, as, unlike $CO_2$, the increased water vapor in the atmosphere does not directly contribute to global warming. It differs in one crucial way from other greenhouse gases: It condenses and falls back to Earth as rain. Nevertheless, the water vapor released in large quantities by combustion engines can play a role by contributing to local climate effects, such as the formation of clouds or fog. Further research is needed to understand the effects. Another question is the influence of liquid water produced during the combustion of fossil fuels, whether through condensation or as part of fine aerosols, for example on sea level rise. Indeed, this could be a crucial discussion for fuel cells. In this particular case, however, the water can be condensed in advance and then released. The fact that the reaction product, $H_2O$, does not need to be stored in the fuel cell is a decisive advantage, as this has a positive effect on energy density, as we will see later in this chapter. The rule of thumb for mobility is therefore "the larger, the more hydrogen."

After this short excursus about the impact of exhaust gases emitted during the burning of fossil fuels and hydrogen, let's return to discussing the energy yield of various elements when used (hypothetically) as fuels in an internal combustion engine. For this purpose, we consider a metal–air battery consisting of a metal anode and an air cathode, assuming that there is an infinite air reservoir and that the reaction product, i.e., the combustion product, is not stored.

*The Periodic System of Elements as the Limiting Factor for Galvanic Cell Design* 9

In the case of lithium, lithium oxide ($Li_2O$) is formed, analogous to the combustion of lithium in an internal combustion engine. The theoretical cell voltage can be calculated as follows:

$$E_{Cell} = -\frac{\Delta G}{z \times F},$$

(1.4)

where
$\Delta G$ is the Gibbs free energy,
$z$ is the number of electrons transferred,
$F$ is the Faraday constant (equal to 96.485 C/mol).

The Gibbs free energy can be easily looked up in standard texts on thermodynamics and physical chemistry [10], as well as in chemical thermodynamics databases [11]. In the case under consideration, the cell voltage is around 2.9 V, and, together with the theoretical specific capacity of lithium of 3.861 mAh/g, as calculated above, equation (1.3) results in a theoretical energy density of 11.197 Wh/kg when lithium is used as fuel in an internal combustion engine. Figure 1.2 and Table 1.1 lists the energy densities of other elements when used as (hypothetical) fuels in an internal combustion engine, which are shown in Figure 1.3. In comparison to gasoline, the oxidation of 1 kg of lithium metal releases 11.197 Wh, not

**Table 1.1.** Theoretical energies of various elements and hexane.

| (Hypoth.) Fuel | Molar mass g/mol | Equiv. number $z$ | Capacity mAh/g | Gibbs energy* kJ/mol | Cell voltage V | Grav. energy density Wh/kg |
|---|---|---|---|---|---|---|
| H | 1.008 | 1 | 26.589 | −237.1 | 1.2 | 31.907 |
| Li | 6.941 | 1 | 3.861 | −561.2 | 2.9 | 11.197 |
| Na | 22.990 | 1 | 1.166 | −375.5 | 2.0 | 3.332 |
| K | 39.098 | 1 | 687 | −322.1 | 1.7 | 1.168 |
| Mg | 24.305 | 2 | 2.206 | −569.3 | 3.0 | 6.618 |
| Ca | 40.078 | 2 | 1.338 | −603.3 | 3.1 | 4.418 |
| Al | 26.982 | 3 | 2.981 | −1582.3 | 1.4 | 4.173 |
| Si | 28.086 | 4 | 3.815 | −850.2 | 2.2 | 8.393 |
| $C_6H_{14}$ | 86.175 | 42 | 13.062 | | 1[†] | 13.062 |

*Note*: *Data taken from Ref. [10]; [†]according to the assumption of a hypothetical fuel cell with small load.

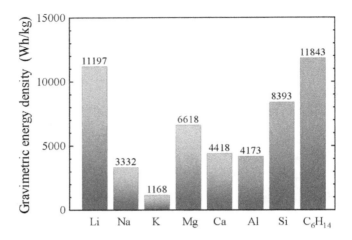

**Figure 1.3.** Theoretical gravimetric energy density of various elements and hexane used as (hypothetical) fuels in an internal combustion engine.

much lower than that of gasoline. Hydrogen produces almost three times this value at 31.907 Wh/kg, making it, by far, the fuel with the highest energy density.

In conclusion, the above thought experiment, in which the various elements are considered as (hypothetical) fuels in an internal combustion engine, demonstrates that the energy yield of lithium approaches that of gasoline. However, it should be emphasized here that the reaction products are released.

Now, let's take a look at the energy density of oxygen batteries and explore the question of what makes the difference.

In a metal–oxygen battery, the energy is stored through the chemical reaction of the metal with oxygen ($O_2$) from the ambient air. The huge difference from the internal combustion engine is that the reaction products are not released into the environment. During discharge, the oxygen combines with the metal ions to form a metal oxide, releasing energy. The oxygen remains bound within the cell, and the cell becomes increasingly heavy. This raises the question of how the energy density changes with the state of charge (SOC)?

Figure 1.4 shows the energy density as a function of the SOC for Li–, Na–, and Zn–$O_2$ batteries. It is clear to see that the energy density decreases with increasing oxygen content during discharge. In total, the

## The Periodic System of Elements as the Limiting Factor for Galvanic Cell Design 11

gravimetric energy density in the case of Li is decreased by around 70% due to the binding of oxygen.

Table 1.2 provides an overview of different types of metal–oxygen batteries, including their reaction mechanisms and calculated gravimetric energy densities. Figure 1.5 shows the calculated energy densities of these batteries in comparison. In the case of Li–O$_2$, the gravimetric energy

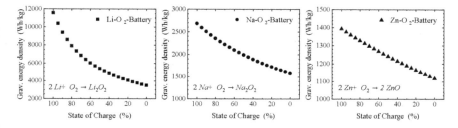

**Figure 1.4.** Theoretical gravimetric energy of various M–O$_2$-batteries (M = Li, Na, Zn) as a function of the SOC.

**Table 1.2.** Theoretical energies of various metal–oxygen batteries.

|  | Total reaction | Molar mass g/mol | Capacity mAh/g | Gibbs energy kJ/mol | Theor. cell voltage V | Grav. energy density Wh/kg |
|---|---|---|---|---|---|---|
| Li–O$_2$ | 2 Li + O$_2$ → Li$_2$O$_2$ | 45.933 | 1.167 | −571.0 | 3.0 | 3.501 |
| Na–O$_2$ | 2 Na + O$_2$ → Na$_2$O$_2$ | 77.978 | 687 | −447.7* | 2.3 | 1.581 |
| Zn–O$_2$ | 2 Zn + O$_2$ → 2 ZnO | 81.418 | 658 | −636.3 | 1.7 | 1.119 |
| Al–O$_2$ | 4/3 Al + O$_2$ + 2 H$_2$O → 4/3 Al(OH)$_3$ | 78.004 | 1.031 | −1.065.7 | 2.8 | 2.886 |
| Mg–O$_2$ | 2 Mg + O$_2$ + 2 H$_2$O → 2 Mg(OH)$_2$ | 58.320 | 919 | −1.195.0 | 3.1 | 2.849 |
| Fe–O$_2$ | 2 Fe + O$_2$ + 2 H$_2$O → 2 Fe(OH)$_2$ | 89.860 | 596 | −499.7 | 1.3 | 775 |
| H$_2$–O$_2$ | 2 H$_2$ + O$_2$ → 2 H$_2$O | 18.015 | 2.975 | −237.1* | 1.2 | 3.571 |

*Note*: *Data taken from Ref. [10]; all other Gibbs energy data taken from Ref. [12].

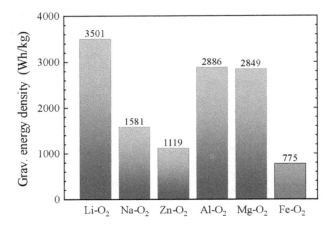

**Figure 1.5.** Theoretical gravimetric energy density of various metal–oxygen batteries.

density is 3501 Wh/kg, which corresponds to only about 30% of the energy yield of lithium as a hypothetical fuel in an internal combustion engine.

## 1.3 What Makes a (Good) Battery?

Back to the initial question: What is the key difference between combustion engines and batteries in terms of energy yield? Unlike internal combustion engines, batteries do not release all their reaction products into the environment. Batteries operate differently, employing chemical reactions to store energy while keeping the components enclosed within the battery. This closed-system operation reduces the energy density of batteries compared to the high energy density of fuels used in internal combustion engines.

However, the closed system of the batteries ensures that no harmful exhaust gases are released into the atmosphere and also allows the processes to be reversible, i.e., the batteries can be recharged. In a nutshell: Eco-friendliness and rechargeability come at a cost – that cost is energy density!

**Excursus – hydrogen fuel cell:** Before we get into this in more detail, let's take a brief look at a special case of the systems under consideration, in which hydrogen is used instead of metal: the $H_2$–$O_2$ cell, i.e., the typical hydrogen fuel cell. The reactions that take place during the operation of a hydrogen fuel cell are listed in Table 1.2. It converts the chemical energy stored in hydrogen and oxygen into electrical energy, producing only water vapor and heat as by-products. In contrast to metal–oxygen batteries, the reaction product is not stored but released directly, which results in the high theoretical energy density of 31.907 Wh/kg (Table 1.1). The continuous supply of hydrogen and oxygen makes the fuel cell a continuously operable energy source.

### 1.3.1 *Rechargeability: A key concept in battery technology*

In terms of what has been developed in batteries over the past two centuries, rechargeability is what they are capable of. It is more about energy throughput than energy density. Think of smartphones and hybrid electric vehicles. Even though energy density seems to be the ruling property, energy throughput is much more important. How could a smartphone or a hybrid electric vehicle operate properly if their batteries were not able to deliver a tremendous energy throughput over their entire lifespan?

Interestingly, if you ask people what makes a bird, many will say: "A bird can fly." But this is not true of all birds. One characteristic that all birds have in common, however, is that they have feathers. The same slight mismatch also applies to modern rechargeable batteries. The pivotal property is not the energy density, but the energy throughput. And, the energy densities of batteries are limited and have limits that cannot be exceeded. This may help in understanding where batteries are useful and where they are not. In this regard, the electromobility powered by batteries is much more of a borderline than a huge breakthrough. The electromobility merely shows the limits of the "rechargeable battery concept."

The following section highlights the crucial importance of the atomic shelf for the reversibility of batteries, including the challenges and progress in this area.

Although air-based batteries are extremely attractive candidates for future energy storage solutions due to their interesting energy densities, they mostly suffer from the reaction products that are formed during discharge. These reaction products are usually insulators that can impede the flow of electrical charge within the battery and make recharging more difficult.

Overpotentials play an important role in the proper functioning of a rechargeable battery, especially during the charging process. In the case of the often-discussed rechargeable Li–air or Li–sulfur batteries, the reaction products, such as $Li_2O_2$ or $Li_2S$, are difficult to split and require significant overpotentials during the charging process. However, these overpotentials reduce the efficiency and the recoverable capacity per cycle.

To ensure the rechargeability of air-based batteries, it is usually necessary to use an electronically conductive matrix made of carbon black or similar materials. In the case of $Li_2O_2$, even expensive catalysts and special electrolytes are necessary to facilitate the electrochemical reactions and improve the reversibility of the battery.

Besides, it is already known from lead–acid batteries that it is extremely challenging to recharge insulating materials, as is the case with $PbSO_4$. $PbSO_4$ is formed during the discharge of the battery, and it forms an impermeable layer on the electrodes that hinders the charging process. This property makes it difficult to fully recharge the battery and thus impairs its overall performance and lifespan.

Therefore, a very smart matrix is required to achieve optimum rechargeability. This matrix plays a crucial role in dealing with insulating reaction products in batteries, as it creates the conditions necessary to effectively overcome them during the charging process and thus ensure the full functionality of the battery.

Let's have a look at the most interesting developments in batteries:

(1) The electrolyte is no longer a part of the reaction (we will follow up on this important insight in Chapter 2).
(2) The electrolyte is only responsible for the transport of ions.
(3) Ions are precisely inserted and extracted at an atomic scale into and from the host matrix of the active materials.
(4) Though many active materials do not have attractive diffusion coefficients for the relevant ions, a second macroscopic heterogeneous matrix of active material particles, binders, a conductive agent, and a liquid electrolyte ("the electrodes") compensate these less attractive diffusion coefficients.
(5) Within the operating window, almost no side reactions take place. This allows the use of very thin materials. As an example, the separator in modern LiBs is about 25 $\mu$m, much thinner than a human hair. At the same time, one can counter-compensate for the rather low conductivities of liquid organic electrolytes in Li-ion cells.

This is what we have called the "shelf." Not obeying rules 1–5 sensitively harms the energy throughput.

And this shelf allows for reducing the overpotentials and quite high power outputs, which are mandatory for applications such as smartphones, hybrid electric vehicles, battery-powered vehicles, and buffer and storage systems.

In this regard, it becomes evident that currently, there are three main technologies ruling the battery world. These are $H^+$, $Li^+$, and $Na^+$ cells. $H^+$ is nothing but nickel–metal hydride. But all these technologies obey rules 1–5.

Violating principles 1–5 results in decreased power capability and worse rechargeability.

Attempts to transfer this approach to multivalent ions, such as $Mg^{2+}$, $Ca^{2+}$, and $Al^{3+}$ are very challenging because these ions are much more immobile due to their multivalent character and are very difficult to insert into and extract from suitable host lattices. Furthermore, there are not yet many known active materials that can host such ions and at the same time have attractive voltage profiles and specific capacities.

Now, since we have elaborated that the shelf is necessary and dominates modern batteries, the story is about calculating the shelf. And there are not more options beyond those in the periodic system of elements! Unlike in the semiconductor industry, where Moore's law describes the continuous improvement and miniaturization of transistors, there is no similar rule in battery technology. Even the lightest ions, such as $Li^+$, offer no guarantee for breaking through the shelf system, on the contrary.

As a rule of thumb, LiBs contain only 3–4 wt.% metallic lithium. Non-active materials, such as electrolytes and current collectors, also play a role and influence the efficiency and capacity of the battery. Please remember that we have a microscopic shelf, namely the active materials or intercalation electrodes, and a macroscopic shelf, as described in rules 1–5. We discuss these matters in more detail in Chapters 2 and 3.

Coming back to make a fair comparison between the hypothetical combustion engines considered in Section 2.2 and LiBs, we need to consider the weight of the inactive materials. Taking into account that LiBs only contain 3–4 wt.% Li, this reduces the gravimetric energy density of a battery to 335–450 Wh/kg.

### 1.3.2 *Future prospects for electrochemical energy storage and conversion*

As a conclusion, modern batteries well address energy throughput, rechargeability, maintenance-free operation, and power capability, as they can effectively reduce overpotentials. These characteristics are crucial for the wide application of batteries in various fields, from portable devices to electric vehicles.

However, if higher energy densities are required, the successor is power-to-X. It can be seen as a battery that is charged once, e.g., using a water electrolyzer. This remains within the framework of the "electrochemical" narrative.

If modern aspects of plasmolysis are also included, such as the splitting of $CO_2$ by a plasma, the narrative of a once-charged battery could be expanded. The much more interesting question is whether this can be accomplished using sustainable power sources, such as photovoltaics and wind.

And a third aspect is the discharge; we also have the heating sector and industry, where hydrogen can simply be burned, for example. The resulting water can then be used again for electrolysis to produce hydrogen.

At the end of the day, a modern battery could be a part of any sustainable cycle, provided it is defined as something that is electrically charged and the resulting product is storable. Even if the product is burned and released, what ultimately matters is whether an entire sustainable cycle can be achieved, such as with real e-fuels having at least a net-zero $CO_2$ emissions balance.

Expanding the battery story, therefore, means manipulating definitions. The rechargeable battery creates a sustainable cycle that can connect different sectors. What is obtained and stored after recharging is a chemical molecule. This seems to be the only way to go to achieve much higher energy densities, as this is the real "game changer" regarding energy densities, as discussed at the beginning of this chapter.

Nevertheless, the "classical" battery will play a very important role in our society, but one in which energy throughput, rechargeability, power capability, and maintenance-free are the most important issues to be addressed directly. The classical fuel cell powered by hydrogen is the bridge between these two worlds and the reason for its outstanding position.

## 1.4 Cost Development and Future Prospects of Energy Storage Technologies and Synthetic Fuels

### 1.4.1 *Development of costs for Li-ion batteries*

This section is dedicated to a detailed analysis of the cost development of LiBs from 2000 to 2030, highlighting the key factors influencing pricing and providing an outlook on future developments.

In 2000, the cost of Li-ion cells was over 2000 €/kWh [13]. Such high costs were one of the biggest obstacles to widespread commercial use, particularly in the field of electromobility. The high prices resulted from the expensive raw materials, the complex manufacturing process, and the limited production capacities. Figure 1.6 shows the global trend of the battery pack price.

Over the years, however, various factors have contributed to drastic cost reductions. Technological innovations play an important role in this. For example, the invention of new electrode materials, such as nickel–manganese–cobalt-oxide (NMC) in 1999 [5], has contributed to an increase in energy density and a reduction in production costs. In addition, automated and advanced manufacturing processes have significantly

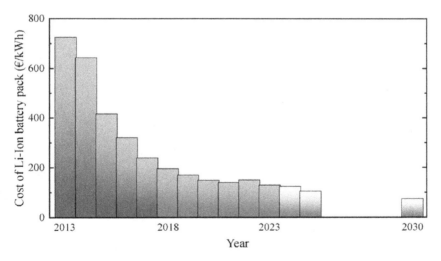

**Figure 1.6.** Worldwide price development of Li-ion battery pack prices from 2010 to 2030.

*Source*: Data taken from Ref. [14].

reduced the production costs. Another reason is the expansion of production capacity due to the increased demand for batteries, which has led to a significant reduction in costs per kWh caused by economies of scale. Optimizing supply chains and ensuring the availability of raw materials helped reduce material costs.

In 2013, the cost of LiB packs was around 725€/kWh, and in 2022, it was just 150 €/kWh [14]. This impressive reduction has significantly accelerated the spread of electric vehicles and renewable energy storage systems.

The trend toward cost reduction continues, and the cost of LiBs is estimated to be around 105 €/kWh in 2025, according to BloombergNEF (BNEF) [14]. The ongoing optimization of production processes, increasing automation, and continuous research into new materials are contributing to this positive development.

BNEF forecasts a price of 75 €/kWh for the period after 2030 [14]. This outlook is based on several factors. The research and introduction of new materials, such as solid-state batteries, which offer higher energy density and greater safety, could further reduce costs. In addition, improved recycling technologies and the integration of a circular economy for battery materials can reduce dependence on expensive raw materials and lower overall costs. Support programs and regulatory frameworks of policies supporting electromobility and renewable energy will also play an important role. The development of the cost of LiB packs, from 725 €/kWh in 2013 to an expected 75 €/kWh after 2030, shows a remarkable reduction of around 90%. This cost reduction is the key to the wider use of batteries and will contribute significantly to the transformation of the energy and mobility sectors.

### 1.4.2 *Costs and development status of e-fuels*

E-fuels, also known as synthetic fuels, are being discussed as an alternative to fossil fuels. They are produced by synthesizing hydrogen and $CO_2$, with the hydrogen being generated by electrolysis using renewable energy. In this section, we examine the current costs and development status of e-fuels and look at their future prospects.

The production costs of e-fuels are currently around 2.20–4.80 €/l [15]. These high costs are mainly due to the energy-intensive production process and the still relatively high prices of renewable energies required for electrolysis. A significant proportion of the production costs is accounted for by the electricity required to produce green hydrogen through electrolysis. The affordability of renewable energy sources such as solar and wind power is steadily improving, directly influencing e-fuel production costs. Countries such as Morocco, Chile, Australia, and China are showing promising advancements in renewable energy production, benefiting from favorable natural conditions and supportive political frameworks. For example, electricity from renewable energy sources is currently available in Morocco at a price of around 4 €ct/kWh. So, while the current cost of e-fuels is still high, there are promising signs that this cost could fall to around 1.20–3.60 €/l by 2050 [15]. This development is largely dependent on the cost of electricity from renewable energy sources.

# References

[1] S. G. Bratsch, Standard electrode potentials and temperature coefficients in water at 298.15 K, *Journal of Physical and Chemical Reference Data*, 18(1), 1–21, 1989.

[2] W. Lee, J. Kim, S. Yun *et al.*, Multiscale factors in designing alkali-ion (Li, Na, and K) transition metal inorganic compounds for next-generation rechargeable batteries, *Energy & Environmental Science*, 13(12), 4406–4449, 2020.

[3] S. Chen, D. Zhao, L. Chen *et al.*, Emerging intercalation cathode materials for multivalent metal-ion batteries: Status and challenges, *Small Structures*, 2(11), 2021.

[4] D. Linden, *Handbook of Batteries*, Fuel and Energy Abstracts, 1995.

[5] H. Zhang, Y. Yang, D. Ren *et al.*, Graphite as anode materials: Fundamental mechanism, recent progress and advances, *Energy Storage Materials*, 36, 147–170, 2021.

[6] Q. Wang, B. Liu, Y. Shen *et al.*, Confronting the challenges in lithium anodes for lithium metal batteries, *Advanced Science*, 8(17), 2101111, 2021.

[7] R. Wang, W. Cui, F. Chu *et al.*, Lithium metal anodes: Present and future, *Journal of Energy Chemistry*, 48, 145–159, 2020.

[8] V. Viswanathan, A. H. Epstein, Y.-M. Chiang *et al.*, The challenges and opportunities of battery-powered flight, *Nature*, 601(7894), 519–525, 2022.

[9] Deutscher Bundestag, *Gesetz zur Änderung des Erdgas-Wärme-Preisbremsengesetzes, zur Änderung des Strompreisbremsegesetzes sowie zur Änderung weiterer energiewirtschaftlicher, umweltrechtlicher und sozialrechtlicher Gesetze*, 26.07.2023.

[10] W. M. Haynes, *CRC Handbook of Chemistry and Physics*, CRC Press, 2014.

[11] T. C. Allison, NIST-JANAF thermochemical tables – SRD 13, 2013.

[12] K. L. Ng, K. Shu, and G. Azimi, A rechargeable Mg|O2 battery, *iScience*, 25(8), 104711, 2022.

[13] M. S. Ziegler and J. E. Trancik, Re-examining rates of lithium-ion battery technology improvement and cost decline, *Energy & Environmental Science*, 14(4), 1635–1651, 2021.

[14] BloombergNEF, Lithium-ion battery pack prices hit record low of $139/kWh, 11/26/2023, https://about.bnef.com/blog/lithium-ion-battery-pack-prices-hit-record-low-of-139-kwh/.

[15] M. Wietschel, P. Plötz, E. Dütschke *et al.*, Diskussionsbeitrag – Eine kritische Diskussion der beschlossenen Maßnahmen zur E-Fuel-Förderung im Modernisierungspaket für Klimaschutz und Planungsbeschleunigung der Bundesregierung vom 28.3.2023.

© 2025 World Scientific Publishing Company
https://doi.org/10.9789811282058_0002

# Chapter 2

# Energy Density vs. Power Density: Implications for Electrolyte Quantity and Filling Procedure

**Johannes Wanner[*] and Kai Peter Birke[†]**

*Fraunhofer IPA, Nobelstrasse 12, Stuttgart, Germany*

[*]*johannes.wanner@ipa.fraunhofer.de*

[†]*kai.peter.birke@ipa.fraunhofer.de*

## 2.1 Comparison of High-Energy and High-Power Cylindrical Battery Cells

High-power battery cells have electrodes specifically designed to facilitate rapid charging and discharging [1, 2]. These electrodes typically feature a larger surface area, which enhances faster ion transfer [3]. Materials such as graphite are commonly used for the anode, while the cathode might utilize lithium iron phosphate (LiFePO4) or other materials that support quick electron movement [4, 5]. In contrast, high-energy battery cells focus on maximizing energy density. The electrodes in high-energy cells are designed to hold more ions, which generally involves a thicker and denser material structure [6, 7]. Common materials for the anode include silicon-based composites, and the cathode often employs nickel manganese cobalt oxide (NMC) or lithium cobalt oxide (LCO) to maximize capacity. To illustrate the differences in properties, a comparison between high-power and high-energy cells is provided in the following [4, 5].

The LG M50 21700 cell is widely used as a high-energy cell in various applications. Chang-Hui Chen *et al.* [5] provides an extensive dataset for and analysis of a cylindrical 21700 commercial cell (LGM50). The LGM50 cell comprises a positive electrode made of NMC 811 and a negative electrode consisting of a bi-component mixture of Graphite-SiOx, with a capacity of 5 Ah.

As a counterexample, consider a high-power LFP cell based on the findings of Prada *et al.* [4]. This cell utilizes a graphite anode and an LFP cathode. In the 21700 format, the resulting cell capacity is 2.85 Ah. The electrode's properties are listed in Table 2.1.

In addition to the capacitance and the height of the jelly roll, only small deviations in the anode properties can be observed, while the cathode's properties exhibit significant variations. The separator has different thicknesses but a similar porosity. The differences in the jelly roll's dimensions and porosity are important parameters for calculating the electrolyte amount in the battery cell.

**Table 2.1.** Cell properties of high-energy and high-power cell.

| Cell | High-energy | High-power |
|---|---|---|
| Source | Chen *et al.* [5] | Prada *et al.* [4] |
| Capacity | 5 Ah | 2.85 Ah |
| Jelly roll height | 65 mm | 60 mm |
| Anode thickness | 100 $\mu m$ | 90 $\mu m$ |
| Anode conductivity | 215 S/m | 215 S/m |
| Anode diffusivity | 3.3e−14 m²/s | 3e−15 m²/s |
| Anode porosity | 25% | 36% |
| Anode grain size | 5.86 $\mu m$ | 5 $\mu m$ |
| Anode current collector thickness | 12 $\mu m$ | 12 $\mu m$ |
| Cathode thickness | 88.8 $\mu m$ | 58 $\mu m$ |
| Cathode conductivity | 0.18 S/m | 0.337 S/m |
| Cathode diffusivity | 4e−15 m²/s | 5.9e−18 m²/s |
| Cathode porosity | 25% | 36% |
| Cathode grain size | 0.522 $\mu m$ | 50 nm |
| Cathode current collector thickness | 16 $\mu m$ | 16 $\mu m$ |
| Separator thickness | 12 $\mu m$ | 25 $\mu m$ |
| Separator porosity | 47% | 45% |

The differences in porosity are significant: The high-energy cell exhibits 25%, which is lower than the 36% found in the high-power cell for both the anode and cathode. Additionally, variations in diffusivity are of particular interest for assessing the electrolyte filling properties of the cell.

The resulting cell design is shown in Figure 2.1. The materials are wound around a central core and inserted into the housing. For ease of handling during winding, there is a cavity in the center. To fit into the housing, there is also a gap between the outer layer and the housing itself. Additionally, cavities are included both below and above the jelly roll to provide space for the tab contacts.

The jelly roll is characterized by an Archimedean spiral shape. In both high-energy and high-power cells, the electrodes are coated on both sides [3]. Specifically, 23 windings are calculated for the high-energy cell, while the high-power cell features 25 windings. This results in an approximate electrode length of 0.85 m for the high-energy cell and 0.94 m for the high-power cell.

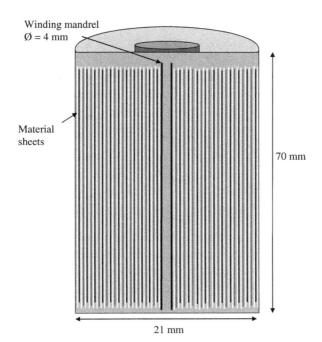

**Figure 2.1.** Cell design of the high-energy cell based on Wanner *et al.* [1].

Table 2.2. Active material volumes based on Refs. [4, 5].

| Cell | High-energy | High-power |
|---|---|---|
| Source | Chen *et al.* [5] | Prada *et al.* [4] |
| Capacity | 5 Ah | 2.85 Ah |
| Jelly roll height | 65 mm | 60 mm |
| Jelly roll inner diameter | 4 mm | 4 mm |
| Jelly roll outer diameter | 19.9 mm | 19.9 mm |
| Jelly roll windings | 23 | 25 |
| Anode pore volume | 1.39 ml | 1.83 ml |
| Cathode pore volume | 1.23 ml | 1.18 ml |
| Separator pore volume | 0.31 ml | 0.63 ml |
| Cell housing height | 68 mm | 68 mm |
| Cell housing inner diameter | 20.2 mm | 20.2 mm |
| Total cavity volume | 2.39 ml | 3.88 ml |
| Total pore volume | 5.87 ml | 7.28 ml |

The resulting volumes are listed in Table 2.2.

Thus, a significant deviation in cavity volume and pore volume is evident due to the electrode design [8]. To accurately calculate the required amount of electrolyte, both the pore volume of a battery and the free volume in the housing must be considered. Ideally, the jelly roll absorbs the entire volume of electrolyte. The high-energy cell weighs 68.38 g. One ml of 1.0 M LiPF6 in the EC/DMC electrolyte weighs 1.3 g [9]. This results in a weight difference of 1.83 g, depending on whether 5.87 ml or 7.28 ml is filled.

## 2.2 Electrode Manufacturing Deviations and Their Influence on Electrolyte Quantity

The manufacturing of battery cells is subject to process variations that impact the parameters mentioned earlier. Therefore, the porosities and electrolyte volumes are assumed to follow a normal distribution for the high-energy cell, shown in Figure 2.2 [1, 10]. According to Peter *et al.* [11], the standard deviations for the porosities and the effective pore radius are small for the separator ($\pm$0.94%), somewhat larger for the anodes ($\pm$1.25%) and considerable for the cathode ($\pm$2%).

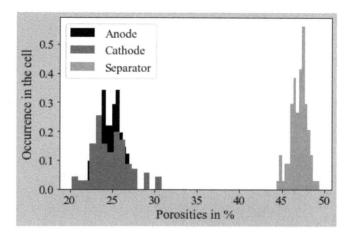

**Figure 2.2.** Porosity fluctuations in electrode and separator manufacturing.

In the worst-case scenario, these fluctuations can result in pore volumes ranging from 4.99 ml to 6.77 ml. Therefore, it is crucial to maintain consistent electrode properties and utilize a higher quantity of electrolyte that can effectively compensate for these variations. Achieving this may involve product modeling integrated into the production process [12].

### 2.2.1 *Product and process modeling for assessing deviations in cell manufacturing*

These models typically represent production variations and are utilized to assess product quality [13, 14]. Reinhart *et al.* [15] introduced an object-oriented methodology in which the battery cell is composed of individual components, each with unique physical properties and functions that affect overall product attributes such as safety and cycle stability, derived from the principal battery class. Kornas *et al.* [16] build upon this concept by advancing the battery model into a multivariate KPI-based framework [16].

The hierarchical structure depicted in Figure 2.3 outlines a multiscale model spanning material, product, and process levels. At the base level, it captures physically measured input characteristics (Cpk), which include material properties, process parameters, and specified targets or tolerances (Cpmk). The second tier details intermediate product characteristics (MCpk), such as cell body thickness, derived from first-level attributes like electrode strip length and winding process properties.

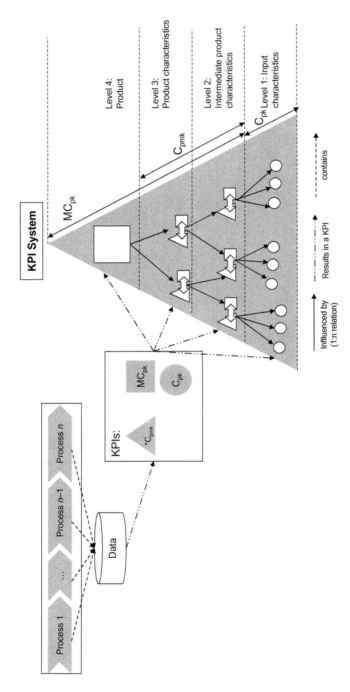

**Figure 2.3.** Structure of a multivariate KPI-based battery model and its use in manufacturing environment. In the style of Kornas et al. [16].

Increased deviations from allowable tolerances at this stage can indicate potential quality losses. The third tier focuses on the final product attributes of the cell, such as Coulomb efficiency or capacity. At the pinnacle of the hierarchy, these properties are aggregated into a singular KPI value. This structured approach facilitates correlation among all input characteristics, aligning them to a unified KPI [16, 17].

## 2.3 Process Modeling of Electrolyte Filling and Wetting

Due to limited accessibility for direct observation of phenomena, inline control to monitor the wetting state during production is currently unfeasible [18]. To ensure uniform and rapid wetting of electrode sheets, methods such as applying vacuum and overpressure during the electrolyte filling process are utilized. Additionally, adjustments to the electrode and electrolyte materials, as well as variations in process temperatures, can influence wetting conditions [19, 20]. However, this limitation hinders the precise quantification of process enhancements. In the literature, various experimental in-situ techniques are used to quantify wetting and address these challenges. Electrical methods like impedance spectroscopy are experimentally employed to assess the wetting level at the cell stage. Macroscopic visualization techniques such as X-ray, neutron radiography, or thermography are used to depict wetting phenomena, although they often involve complex setups and provide insufficient temporal and spatial resolution of wetting processes [18]. An alternative approach, the wetting balance test, simplifies the complexity of assessing wetting by focusing on specific test objects and allows for gravimetric or visual analysis of wetting progress in the electrode. This visual assessment method enables measurement and quantification of electrolyte movements on a smaller scale [10, 21].

In addition to experimental approaches, the literature discusses several numerical fluid dynamics simulations, particularly utilizing the lattice Boltzmann method (LBM), to explore wetting phenomena at the pore scale [22]. The LBM technique is frequently applied to fluid simulations in porous media and intricate geometries. In particular, the multicomponent Shan–Chen pseudopotential method facilitates the modeling of fluid–fluid and solid–fluid interaction forces, interfacial tensions, and adhesion forces relevant to wetting phenomena. While the LBM has demonstrated its efficiency in fields such as geoscience and fuel cell technology, its applications in the study of electrolyte transport phenomena in battery materials have been less common. Notably, researchers including Jeon and colleagues, Lee et al., and Mohammadian and Zhang have employed LBM simulations to examine how structural characteristics of electrodes influence filling processes on a two-dimensional scale, often simplifying electrode geometries significantly [22–24].

Sauter et al. [25] employed a binarized three-dimensional microstructure of a separator derived from focused ion beam scanning electron microscopic tomography data. Similarly, Lautenschlaeger et al. [22] used realistic 3D reconstructions of cathodes, including binders, based on binarized CT images, while Shin et al. [27] reconstructed PE separators using FIB-SEM tomography image processing. Shodiev et al. [20] utilized actual tomography data for cathode visualization and stochastically generated semi-realistic microstructures for the anode and separator. Hagemeister et al. [26] applied a multiscale commercial computational fluid dynamics program to assess how process parameters affect the filling process, using 3D electrode geometries of the cathode reconstructed from micro-CT scan images. While most studies focus on simulating wetting for individual electrode geometries, researchers such as Lautenschlaeger et al. [22], Shin et al. [27], and Shodiev et al. [20] also explored the impacts at the interfaces between electrodes and separators.

Understanding these process phenomena, coupled with advanced sensor technology and precise modeling, enables the creation of a Digital Twin of the wetting process. This digital representation can assist in defining specific process timings for each instance [17, 28].

## 2.4 Capillary Wetting of Battery Materials

Regardless of the material type, the Lucas–Washburn equation effectively describes capillary rise in porous media [29]. In the context of battery materials, which are porous, the wetting process is primarily driven by capillary pressure $P_C$. Additionally, the influence of viscosity reduction, as dictated by Darcy's law $\nabla P_V$, must be considered [30]. Factors such as hydrostatic pressure $P_H$ and ambient pressure $p_0$ also play significant roles during the time frame in which capillary effects are observed:

$$\frac{dh}{dt} = \frac{\sum P}{8h\mu_L}(r_{\text{eff}}^2), \quad \text{with} \quad \sum P = P_C - P_H - \nabla P_V - p_0. \quad (2.1)$$

In this context, the height of liquid penetration is represented by $h$ and $\mu_L$ indicates the dynamic viscosity of the liquid. The specific porous media, characterized by variable pore sizes, are described using an effective pore radius distribution $r_{\text{eff}}$. The capillary pressure $P_C$ is defined as follows:

$$P_C = \frac{2\gamma \cos(\theta)}{r_{\text{eff}}}.$$ (2.2)

Capillary wetting is influenced by the surface tension between the liquid and the gas, denoted as $\gamma$, and the contact angle between the liquid and the solid surface, represented by $\theta$. The hydrostatic pressure of the liquid is defined as follows:

$$P_H = g \cdot \rho L \cdot h \cdot \sin(\psi).$$ (2.3)

In this setup, $\rho$ represents the density of the fluid in question, g stands for gravity, and $\psi$ indicates the alignment angle of the porous media relative to a horizontal axis. The viscous pressure loss $\nabla P_V$ of the liquid corresponds to the following:

$$\nabla P_V = \frac{\mu_L \cdot h \cdot dh/dt}{k},$$ (2.4)

where $k$ describes the permeability. Here, the atmospheric pressure and air in the pores are neglected, reducing the formula to the following:

$$\frac{dh}{dt} = \frac{r_{\text{eff}}^2 k}{h\left(8k\,\mu_E + r_{\text{eff}}^2\,\mu_E\right)}\left[\frac{2\gamma \cos(\theta)}{r_{\text{eff}}} - \rho_E \cdot g \cdot h \cdot \sin(\psi)\right].$$ (2.5)

In a cylindrical cell, the winding process significantly influences the composition of the porous medium and, consequently, the wetting speed of the individual materials. Therefore, it is essential to consider the winding process when calculating the wetting height in cylindrical cells. During winding, a tensile force is applied, starting at $F = 10$ N and decreasing linearly by 0.36 N with each winding, as documented in other work [3, 31]. Given that the compression modulus of graphite is substantially higher than that of the separator, it is presumed that only the separator undergoes compression [32]. The resulting winding stress and the reduction in separator thickness are shown in Figure 2.4.

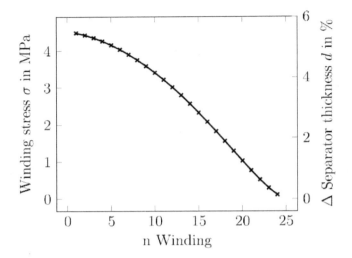

**Figure 2.4.** Winding stress as a function of the 24 windings of a 21700 cylindrical cell, based on Wanner et al. [31].

In summary, the wetting process of a compressed material stack is illustrated in Figure 2.5, where wetting primarily occurs at the transition between materials. This is attributed to the larger pore sizes and increased permeability at these transition points. Subsequently, wetting proceeds horizontally based on the wetting height established at these transitions. The wetting speed, resulting pore radii, and permeability at the material transitions and the separator are significantly influenced by the external pressure applied during compression in either the jelly roll or the pouch cell [31].

Most of the electrolyte properties are well known from manufacturing data or can be measured using a viscometer. An optical contact angle measurement and contour analysis system can determine the contact angle between the electrolyte and the materials.

### 2.4.1 *Identification of pore size and permeability of materials using lattice Boltzmann simulation*

The primary parameter influencing capillary wetting is the pore size of the material. Simulation methods, such as the lattice Boltzmann simulation, enable the determination of the overall pore size distribution, encompassing

*Energy Density vs. Power Density* 31

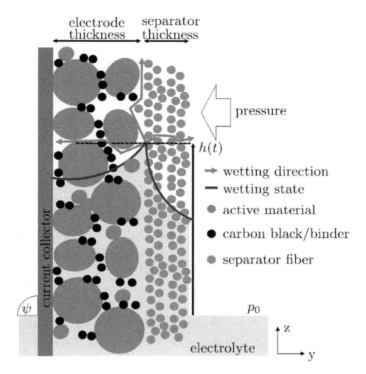

**Figure 2.5.** Illustration of a wetting process at the material transition, following the style of Wanner *et al.* [31].

the smaller pores that are significant for the complete wetting of the materials. To generate realistic simulation geometry, 3D surface profile microscopy images are taken of the materials. The surface height of the material is used to create a high-resolution binary 3D model of the surface. The electrodes are processed separately and combined with the separator. In this process, the 3D model of the electrode is merged with the model of the separator, aligning the surfaces toward each other. The separator's surface structure is soft and contains a few large fibers, making a simple addition of the geometries inaccurate. Therefore, the transition is realized as a logical and of the upper layers of both geometries [21, 31].

Each of the geometries is individually processed and paired with the same separator before being integrated into the lattice Boltzmann framework. The wetting process is carried out in the $y$-direction, with pressure boundary conditions established at both the inlet and outlet in this direction. A bounce-back function is utilized for the opposing media to prevent backflow. Initially, the geometry's pores are filled with air. Both the electrodes and the separator are equipped with a bounce-back mechanism. The contact angles for each material are incorporated using the adhesion parameter for the electrolyte. The wetting process relies on the pressure differential between the fluid phases, governed by the interfacial tension of the interaction parameter $G_C$ and the time step $\Delta t$. After an initialization period of 500 iterations without any external forces, the fluid density along the x-axis is assessed, allowing for the calculation of the pressure difference [21, 31]. Using equation (2.2), the pore sizes at respective positions in the geometry can be calculated. The result is shown in Figure 2.6.

The relative permeability $k$ represents the ability of a fluid to flow through a porous medium. This concept is included in the analytical Darcy equation (2.4), which calculates viscous pressure loss for liquids or gases. To find the relative permeability of a porous medium, the lattice Boltzmann setup is filled with a fluid and an external force is applied in the x-direction. By measuring the average flow velocity through the pores

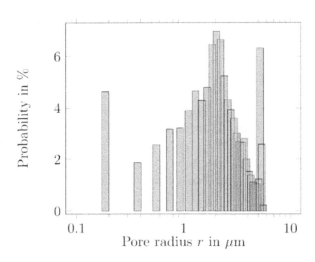

**Figure 2.6.** Pore size distribution determination by lattice Boltzmann simulation, following the style of Wanner et al. [1, 31].

in the $x$-direction and using the kinetic viscosity of the fluid, the relative permeability can be calculated [21, 31].

Based on Fries *et al.* [30], the porosity $\phi$ of the porous medium can be determined from the pore size and permeability:

$$\phi = \frac{8k}{r_{eff}^2}. \tag{2.6}$$

## 2.4.2 *Replication of the wetting height*

Using this lattice Boltzmann simulation method, taking the median of the pore sizes, the permeability, and the measured values of the contact angle, the results are in Table 2.3.

For the electrolyte, the commonly used LP572 (BASF, Germany) is assumed: A 1 M solution of LiPF6 in a mixture of ethylene carbonate (EC) and ethyl methyl carbonate (EMC with a weight ratio of EC:EMC of 3:7), and containing 2 wt.% vinylene carbonate (VC). At 20°C, the density is $\rho = 1201$ kg/m³, the surface tension $\gamma = 42.2$ mN/m, and the dynamic viscosity $\mu_L = 2.3$ mPa s [9].

**Table 2.3.** Pore radius, permeability, and contact angle of the high-energy and high-power cells.

| Cell | High-energy | High-power |
|---|:---:|:---:|
| Anode pore radius | 500 nm | 500 nm |
| Anode porosity | 25% | 36% |
| Anode permeability | 7.81 e−15 m² | 1.13 e−14 m² |
| Anode contact angle | 16.8° | 16.8° |
| Cathode pore radius | 300 nm | 300 nm |
| Cathode porosity | 25% | 36% |
| Cathode permeability | 2.81 e−15 m² | 4.05 e−15 m² |
| Cathode contact angle | 16.8° | 16.8° |
| Separator pore radius | 200 nm | 200 nm |
| Separator porosity | 47% | 45% |
| Separator permeability | 2.35 e−15 m² | 2.25 e−15 m² |
| Separator contact angle | 30° | 30° |

The wetting process in a high-energy cell exhibits notable differences. The anode wets the entire 65 mm height of the material in just 2.8 hours. In contrast, the separator and the cathode wet more slowly, taking 4.6 hours to reach the same height shown in Figure 2.7.

This discrepancy is even more apparent in the high-power cell. Here, the anode wets up to a height of 60 mm in just 1.7 hours, followed by the cathode at 2.9 hours, and finally the separator at 4 hours shown in Figure 2.8.

In both cases, the relatively thin separator can also be horizontally wetted from the anode or cathode, but horizontal wetting of the cathode starting from the anode through the separator is unlikely. Consequently, the wetting time is primarily constrained by either the separator or the cathode. In the case of the high-energy cell, this wetting process takes 4.6 hours, while for the high-power cell, it is 2.9 hours, despite the electrode height being 5 mm lower in the latter case. For comparison, the wetting heights of the cathode are illustrated again in Figure 2.9.

Essentially, the discrepancy highlights the differences in porosity and, consequently, the permeability of the materials. The same pore radius was intentionally chosen for both the high-power and high-energy cells to emphasize this distinction.

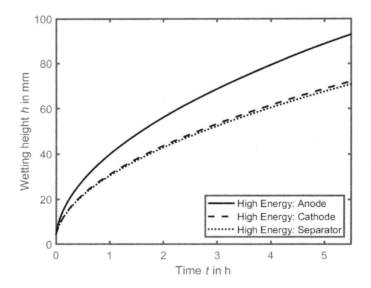

**Figure 2.7.** Wetting height of the high-energy battery cell materials.

*Energy Density vs. Power Density* 35

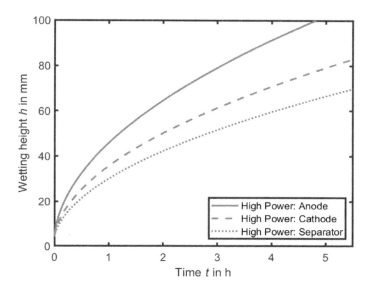

**Figure 2.8.** Wetting height of the high-power battery cell materials.

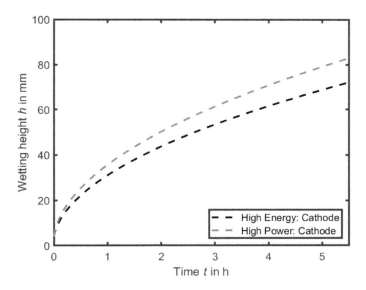

**Figure 2.9.** Comparison of cathode wetting between high-energy and high-power cells.

## 2.5 Influence of the Pore Structure on Conductivity

Modern battery cells, including NiMH, Li, and Na-ion cells, have unique microscopic structures in their electrodes, as shown in Figure 2.5.

A matrix of conductivity enhancers and electrolytes enables a significantly improved effective contact area. The principle of maximizing surface area is akin to the intricate coastline of Norway when considering all its fjords. Ions and electrons recombine on the surfaces of active material particles, crucial for subsequent solid-state diffusion, where particle size plays a critical role. LFP (lithium iron phosphate), for instance, is inherently a very poor electronic conductor, with diffusion occurring predominantly in one dimension. Carbon coating reduces particle size distribution to below 1 $\mu$m, enabling the use of LFP in battery applications and showcasing its current properties.

This approach minimizes side reactions with Li-ions and Na-ions within the operating range, enabling the use of ultra-thin electrodes, small particle sizes, and separators (thinner than a human hair). This is made possible by the ability to mass-produce these thin components with high quality and reproducibility. Although the active materials in the electrodes may have less favorable diffusion coefficients and organic electrolytes show significantly reduced conductivities compared to inorganic aqueous electrolytes (approximately two orders of magnitude lower), it is these unique microscopic conditions that enable the exceptional energy density and performance of Li-ion cells.

However, if this special microscopic structure is not maintained or can no longer be sustained for certain reasons (e.g., the use of metallic Li-electrode or solid-state batteries), the advantageous conditions can change drastically. This consideration is crucial when discussing alternative battery technologies.

## 2.6 Conclusion and Outlook

In summary, the wetting process in a compressed material stack is primarily characterized by wetting occurring at the material transitions. This is facilitated by the larger pore sizes and increased permeability found at these transition points. Horizontal wetting follows vertical wetting, and the speed of wetting is influenced by the resulting pore radii, permeability, and porosity of the respective materials.

In addition to the cell design, it is crucial to document this information and process variations in the production process to accurately calculate deviations in the wetting time of materials. Moving forward, these data can be utilized to develop a digital twin of the wetting process, which would aid in predicting the precise wetting times for the electrodes used in the cell.

# References

[1] J. Wanner, M. Weeber, K. P. Birke *et al.*, Potentials of a Digital Twin implementation in the wetting process in battery cell manufacturing, *Procedia CIRP*, 118, 987–992, 2023.

[2] T. Waldmann, R.-G. Scurtu, D. Brändle *et al.*, Effects of tab design in 21700 Li-ion cells: Improvements of cell impedance, rate capability, and cycling aging, *Energy Technology*, 11(5), 2200583, 2023.

[3] S. Baazouzi, N. Feistel, J. Wanner *et al.*, Design, properties, and manufacturing of cylindrical Li-ion battery cells: A generic overview, *Batteries*, 9(6), 309, 2023.

[4] E. Prada, D. Di Domenico, Y. Creff *et al.*, A simplified electrochemical and thermal aging model of LiFePO 4-graphite Li-ion batteries: Power and capacity fade simulations, *Journal of the Electrochemical Society*, 160(4), A616–A628, 2013.

[5] C.-H. Chen, F. Brosa Planella, K. O'Regan *et al.*, Development of experimental techniques for parameterization of multi-scale lithium-ion battery models, *Journal of the Electrochemical Society*, 167(8), 80534, 2020.

[6] T. Waldmann, R.-G. Scurtu, K. Richter *et al.*, 18650 vs. 21700 Li-ion cells: A direct comparison of electrochemical, thermal, and geometrical properties, *Journal of Power Sources*, 472, 228614, 2020.

[7] J. B. Quinn, T. Waldmann, K. Richter *et al.*, Energy density of cylindrical Li-ion cells: A comparison of commercial 18650 to the 21700 cells, *Journal of the Electrochemical Society*, 165(14), A3284–A3291, 2018.

[8] J. Full, J. Wanner, S. Kiemel *et al.*, Comparing technical criteria of various lithium-ion battery cell formats for deriving respective market potentials, in *2020 IEEE Electric Power and Energy Conference (EPEC)*, pp. 1–6, IEEE, Piscataway, NJ, 2020.

[9] Merck KGaA, Lithiumhexafluorophosphat-Lösung, https://www.sigmaaldrich.com/DE/de/product/aldrich/746711.

[10] J. Wanner and K. P. Birke, Comparison of an experimental electrolyte wetting of a lithium-ion battery anode and separator by a lattice Boltzmann simulation, *Batteries*, 8(12), 277, 2022.

[11] C. Peter, K. Nikolowski, S. Reuber *et al.*, Chronoamperometry as an electrochemical *in situ* approach to investigate the electrolyte wetting process of lithium-ion cells, *Journal of Applied Electrochemistry*, 50(3), 295–309, 2020.

[12] J. Wanner, J. Bahr, J. Full *et al.*, Technology assessment for digitalization in battery cell manufacturing, *Procedia CIRP*, 99, 520–525, 2021.

[13] S. Singh, J. Wanner, and M. Weeber, Modeling and simulation in battery cell manufacturing – concepts and applications, in P. Birke, M. Weeber, and M. Oberle (Eds.), *Handbook on Smart Battery Cell Manufacturing: The Power of Digitalization*, pp. 75–119, World Scientific, New Jersey, 2022.

[14] J. Wanner, M. Weeber, K. P. Birke *et al.*, Quality modelling in battery cell manufacturing using soft sensing and sensor fusion – a review, in *2019 9th International Electric Drives Production Conference (E|DPC)*: 3 and 4 December 2019, Esslingen, Germany: Proceedings, pp. 1–9, IEEE, Piscataway, NJ, 2019.

[15] G. Reinhart, J. Kurfer, M. Westermeier *et al.*, Integrated product and process model for production system design and quality assurance for EV battery cells, *Advanced Materials Research*, 907, 365–378, 2014.

[16] T. Kornas, E. Knak, R. Daub *et al.*, A multivariate KPI-based method for quality assurance in lithium-ion-battery production, *Procedia CIRP*, 81, 75–80, 2019.

[17] P. Birke, M. Weeber, and M. Oberle (Eds.), *Handbook on Smart Battery Cell Manufacturing: The Power of Digitalization*, World Scientific, New Jersey, 2022.

[18] N. Kaden, R. Schlimbach, Á. Rohde García *et al.*, A systematic literature analysis on electrolyte filling and wetting in lithium-ion battery production, *Batteries*, 9(3), 164, 2023.

[19] Y. Sheng, C. R. Fell, Y. K. Son *et al.*, Effect of calendering on electrode wettability in lithium-ion batteries, *Frontiers in Energy Research*, 2, 121653, 2014.

[20] A. Shodiev, E. Primo, O. Arcelus *et al.*, Insight on electrolyte infiltration of lithium ion battery electrodes by means of a new three-dimensional-resolved lattice Boltzmann model, *Energy Storage Materials*, 38, 80–92, 2021.

[21] J. Wanner and K. P. Birke, Investigation of the influence of electrode surface structures on wettability after electrolyte filling based on experiments and a lattice Boltzmann simulation, *Energies*, 16(15), 5640, 2023.

[22] M. P. Lautenschlaeger, B. Prifling, B. Kellers *et al.*, Understanding electrolyte filling of lithium-ion battery electrodes on the pore scale using the lattice Boltzmann method, *Batteries & Supercaps*, 5(7), 1–14, 2022.

[23] D. H. Jeon, Enhancing electrode wettability in lithium-ion battery via particle-size ratio control, *Applied Materials Today*, 22, 100976, 2021.

[24] S. K. Mohammadian and Y. Zhang, Improving wettability and preventing Li-ion batteries from thermal runaway using microchannels, *International Journal of Heat and Mass Transfer*, 118, 911–918, 2018.

[25]   C. Sauter, R. Zahn, and V. Wood, Understanding electrolyte infilling of lithium ion batteries, *Journal of the Electrochemical Society*, 167(10), 100546, 2020.

[26]   J. Hagemeister, F. J. Günter, T. Rinner *et al.*, Numerical models of the electrolyte filling process of lithium-ion batteries to accelerate and improve the process and cell design, *Batteries*, 8(10), 159, 2022.

[27]   S. Shin, H. Kim, T. Maiyalagan *et al.*, Sophisticated 3D microstructural reconstruction for numerical analysis of electrolyte imbibition in Li-ion battery separator and anode, *Materials Science and Engineering: B*, 284, 115878, 2022.

[28]   J. Wanner, M. Weeber, K. P. Birke *et al.*, Production planning and process optimization of a cell finishing process in battery cell manufacturing, *Procedia CIRP*, 112, 507–512, 2022.

[29]   E. W. Washburn, The dynamics of capillary flow, *Physical Review*, 17(3), 273–283, 1921.

[30]   N. Fries and M. Dreyer, An analytic solution of capillary rise restrained by gravity, *Journal of Colloid and Interface Science*, 320(1), 259–263, 2008.

[31]   J. Wanner and K. P. Birke, Investigation of the influence of cell compression after winding on electrolyte wettability based on experiments and lattice Boltzmann simulation, *Journal of Energy Storage*, 87, 111410, 2024.

[32]   H.-S. Chen, S. Yang, W.-L. Song *et al.*, Quantificational 4D visualization and mechanism analysis of inhomogeneous electrolyte wetting, *eTransportation*, 16, 100232, 2023.

© 2025 World Scientific Publishing Company
https://doi.org/10.9789811282058_0003

# Chapter 3

# The Best Battery Cell Candidates for Highest Energy Density

**Kathrin Schad[*] and Kai Peter Birke[†]**

*Fraunhofer IPA, Nobelstrasse 12, Stuttgart, Germany*

*[*]kathrin.schad@ipa.fraunhofer.de*

*[†]kai.peter.birke@ipa.fraunhofer.de*

## 3.1 Introduction

Chapters 1 and 2 extensively discussed the importance of both the microscopic and macroscopic shelves as crucial prerequisites for modern batteries, focusing on energy throughput, power capability, rechargeability, and maintenance-free operation.

However, this discussion remained more qualitative than quantitative. In this chapter, we shift our focus to a more quantitative perspective. While some battery approaches may appear remarkable and promising, they may fall short when considering the optimal microscopic and macroscopic shelves.

We explore the most prominently discussed battery cell chemistries: lithium-ion, lithium–metal, all-solid-state, lithium–sulfur, lithium–air, sodium-ion, zinc–, and lead–acid batteries. Each of these options touts potential for high energy densities or attractive cost (€/kWh). However, the distribution of the bill of materials plays a crucial role in forming a comprehensive and accurate assessment of these technologies. Expected increases in energy densities for automotive applications or significant cost reductions for stationary battery applications can sometimes prove elusive or unrealistic.

In this chapter, we introduce a flexible calculation blueprint that considers the impact of passive cell components on gravimetric energy density across various cell chemistries.

The starting point is always the chemical reaction within a battery cell, centered solely on the active materials, as intensively discussed in Chapter 1. This reaction equation can be easily converted into g/mol once the stoichiometric ratios are established. We then formally introduce additional terms, referred to here as BIRKE-summands, on both sides of the chemical equation based on the active electrode materials. This formal inclusion enables a direct recalculation of the energy density. The challenge lies in implementing these terms effectively.

Next, we demonstrate through a use case for each of the examined cell chemistries how passive, non-electrochemically active components formally impact the energy density of the battery cell, leading to its decrease.

Starting with the standard lithium-ion battery, we transfer these insights to lithium–metal-based cells. We demonstrate that the additional weight of solid-state electrolytes (SSEs) can significantly impact energy density. In the case of lithium–sulfur or lithium–air batteries, the substantial weight influence of increased conductive additive or catalyst content in the cathode becomes evident. Transitioning to sodium-ion batteries, it becomes apparent that energy density is also largely determined by the selection of active and passive cell components. Similarly, in zinc batteries, emphasis is placed on the influence of the heavy electrolyte and zinc anode. Lastly, we analyze the lead–acid battery, highlighting the weight contribution of its high electrolyte demand.

## 3.2 Achievable Energy Density

This section begins by introducing and discussing the most prominent cell chemistries currently under discussion. These selected cell chemistries are evaluated solely based on their electrochemically active cell components to calculate the theoretical energy density $E_{th}$. Introducing a cell-chemistry-flexible calculation blueprint that incorporates passive cell components into the energy density equation yields an intermediate energy $E_{inter}$. This calculated $E_{inter}$ serves as a primary outcome of this chapter and a crucial

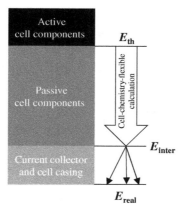

**Figure 3.1.** Schematic illustration of the energy density calculation path showing the theoretical energy density $E_{th}$, the intermediate energy density $E_{inter}$, and the real energy density $E_{real}$, depending on the considered cell components.

step toward determining the actual energy density of a battery cell, $E_{real}$, as illustrated in Figure 3.1.

The concept of passive cell components and their integration within the cell follows consistent principles across all considered cell chemistries, enabling the development of a cell-chemistry-flexible calculation approach to determine $E_{inter}$. However, the high variability in current collector designs and individual cell casing concepts poses challenges for direct comparisons between different cell chemistries and the application of a standardized calculation approach, as illustrated by the varying arrows on the path to $E_{real}$ in Figure 3.1. To enhance comparability, this chapter exclusively focuses on determining $E_{inter}$, thereby excluding the influence of current collectors and casings on energy density in the calculations.

### 3.2.1 *Cell chemistries and their active cell components*

In the following section, we briefly introduce the most widely discussed cell chemistries along with their respective pros and cons. It is important to note that this classification is based exclusively on the electrochemically active cell components, hereafter referred to as active cell components. Consequently, ASSBs, for example, are categorized under lithium–metal batteries. Further subdivision into additional categories will be discussed later in the chapter, once passive cell components are introduced:

(1) Lithium-ion battery (LiB)
(2) Lithium–metal (LiM) battery
(3) Lithium–sulfur (Li–S) battery
(4) Lithium–air (Li–air) battery
(5) Sodium-ion battery (SiB)
(6) Zinc battery (ZB)
(7) Lead–acid (Pb) battery

Figure 3.2 illustrates schematic cell compositions based solely on their active cell components, highlighting prominent representatives of

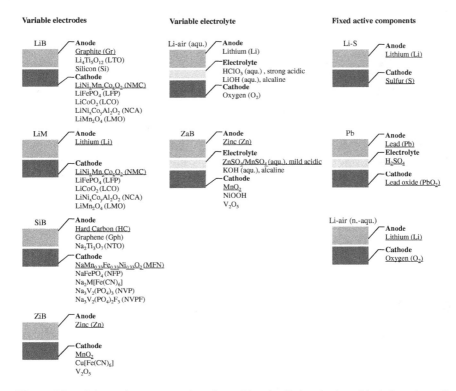

**Figure 3.2.** Schematic representation of considered cell chemistries with their active cell components, categorized by fixed or variable characteristics.

these materials. The examples chosen in this chapter for calculating the energy density $E_{inter}$ are underscored. The various cell chemistries are categorized based on whether their active cell components are fixed or variable within the cell chemistry.

The following sections delve into the specific cell chemistries under consideration. Emphasis is placed on their operational principles and active cell components, which define their theoretical energy density $E_{th}$. Additionally, the advantages and disadvantages of each cell technology are discussed, underscoring the necessity for passive cell components.

## (1) Lithium-ion battery

The standard LiB stands out as a top performer due to its efficient ion storage capabilities at both microscopic and macroscopic levels, which were recognized with the Nobel Prize in 2019. However, a significant challenge remains: achieving substantial increases in energy density without compromising its fundamental design principles. This advancement appears feasible only through modifications to these principles while maintaining adequate performance and lifespan. Additionally, the drive toward highly cost-effective and sustainable battery technologies highlights certain limitations of LiBs in these domains. Consequently, our discussion also explores alternative cell chemistries.

Figure 3.2 illustrates the absence of a standardized setup for LiB in terms of the active materials employed. Depending on the material, lithium storage in the anode can be categorized into three types: intercalation (e.g., graphite (Gr), $Li_4Ti_5O_{12}$ (LTO)), conversion, and alloy formation (e.g., silicon), with graphite being the most widely used anode material. The selection of active cathode materials is similarly diverse, primarily featuring lithium metal oxides such as lithium nickel manganese cobalt oxide ($LiNi_0.8Co_{0.1}Mn_{0.1}O_2$, NMC811), lithium iron phosphate ($LiFePO_4$, LFP), lithium cobalt oxide ($LiCoO_2$, LCO), lithium nickel cobalt aluminum oxides ($LiNi_xCo_yAl_zO_2$, NCA), and lithium manganese oxide ($LiMn_2O_4$, LMO). For the exemplary energy density calculation, NMC811 was chosen as the cathode active material and (Gr) was used as the anode active material. The resulting discharge reaction equation and cell voltage $U$ are presented in Table 3.1(1):

$$E_{th} = \frac{U \, z \, F}{M_{discharge\ products}} \tag{3.1}$$

The theoretical energy density $E_{th}$ of an LiB is 372 Wh/kg. Here, $M_{discharge\ products}$ represents the molar mass of the discharge products, $F$ (26801 Ah/mol) denotes the Faraday constant, and $z$ indicates the number of electrons transferred during the reaction.

### (2) Lithium–metal battery

According to Figure 3.2, LiM batteries utilize lithium metal as the anode with a variety of possible cathode materials, similar to LiBs. Lithium metal boasts a theoretical capacity of 3860 mAh/g, significantly surpassing graphite's theoretical capacity of 360 mAh/g [1]. Moreover, employing lithium metal enables achieving a minimum anode voltage of 0.0 V versus $Li/Li^+$, compared to approximately 0.1 V versus $Li/Li^+$ for (Gr), thereby increasing the overall cell voltage to 3.8 V. When utilizing the same cathode active material NMC811 as in the LiB example calculation, the theoretically achievable energy density $E_{th}$ reaches 523 Wh/kg (refer to Table 3.1(2)).

However, the actual energy density advantage at the cell level is diminished by several factors. The need for sacrificial or excess lithium to compensate for initial losses, as well as unresolved cycling stability issues with liquid electrolytes, which also pose safety risks, are significant concerns. These challenges have spurred interest in solid electrolyte batteries, or all-solid-state batteries (ASSBs), which promise substantial advancements in energy density and battery safety. Over the past decades, SSEs have undergone remarkable development, particularly in enhancing ionic conductivities. In the 1980s, the best average room temperature conductivities of Li-ion conducting SSEs were about $10^{-6}$ S/cm. Today, conductivities of $10^{-3}$ S/cm or higher are achievable. Despite these advancements, several critical aspects have received less attention, including stability against electrodes, sensitivity to humidity, side reactions, and, most importantly, the impact on the gravimetric energy density of the battery cell. ASSBs require high pressure and operating temperatures of around 50–60°C to achieve notable results, which have only been validated over a few hundred cycles. An additional issue with ASSBs is the need for robust casing to maintain pressure, as contact losses in the heterogeneous structure can occur without sustained pressure. Cell production under pressure is necessary, making ASSB batteries viable primarily through the use of significantly thinner layers, known as thin-film batteries. However, these are difficult to manufacture, requiring a substrate and involving

**Table 3.1.** Discharge reaction, cell voltage $U$, and theoretical energy density $E_{th}$ of selected cell chemistries.

| | System | Anode | Electrolyte | Cathode | Discharge reaction | Cell voltage (V) | $E_{th}$ (Wh/ kg) |
|---|---|---|---|---|---|---|---|
| (1) | LiB | $LiC_6$ | | NMC811 | $LiC_6 + 2\ Li_{0.5}Ni_{0.8}Co_{0.1}Mn_{0.1}O_2 \rightarrow C_6 + 2\ LiNi_{0.8}Co_{0.1}Mn_{0.1}O_2$ | 3.7 [1] | 372 |
| (2) | LiM | Li | | NMC811 | $0{,}5\ Li + Li_{0.5}Ni_{0.8}Co_{0.1}Mn_{0.1}O_2 \rightarrow LiNi_{0.8}Co_{0.1}Mn_{0.1}O_2$ | 3.8 [1] | 523 |
| (3) | Li–S | Li | | S | $2\ Li + S \rightarrow Li_2S$ | 2.2 [2] | 2567 |
| (4) | Li–air | Li | | $O_2$ | $2Li + O_2 \rightarrow Li_2O_2$ | 3.0 [2] | 3505 |
| (5) | SiB | HC | | MFN | $NaC_6 + 2\ Na_{0.5}Mn_{0.33}Fe_{0.33}Ni_{0.33}O_2 \rightarrow C_6 + 2\ NaMn_{0.33}Fe_{0.33}Ni_{0.33}O_2^{1,\,2}$ | 3.0 [3] | 274 |
| (6a) | ZiB | Zn | | $MnO_2$ | $Zn + 2\ MnO_2 \rightarrow ZnMn_2O_4$ | 1.3 [4] | 291 |
| (6b) | ZaB | Zn | 1M $ZnSO_4$ 1M $MnSO_4$ (aq.) mild acid | $MnO_2$ | $Zn + MnO_2 + 2\ H_2SO_4 \rightarrow Mn^{2+} + 2H_2O + Zn^{2+} + 2SO_4^{2-}$ | 1.5 [5] | 231 |
| (7) | Pb | Pb | $H_2SO_4$ (aq.) | $PbO_2$ | $Pb + PbO_2 + 2\ H_2SO_4 \rightarrow 2\ PbSO_4 + 2\ H_2O$ | 2.0 [6] | 167 |

*Notes*: [1]The reaction equation of SiB makes a simplified assumption that HC behaves like $C_6$. [2]The reaction equation of SiB makes a simplified assumption that MFN behaves similarly to NMC811 in terms of desodiation depth.

costly production processes. Alternatives such as silicon offer a theoretical capacity of approximately 3800 mAh/g, similarly to lithium metal. Nevertheless, formation losses and rapid aging over cycles can negate the energy density gains at the cell level.

However, there is considerable excitement surrounding the prospect of enhancing energy density using solid electrolytes. This excitement primarily stems from the potential to employ metallic lithium instead of conventional negative electrodes, a development that shows substantial promise. In summary, while materials such as lithium metal and silicon demonstrate impressive theoretical capacities, the practical challenges of integrating them into stable, high-energy-density cells are significant. Understanding and addressing these challenges are crucial for advancing battery technology beyond the current limitations of LiBs. The primary challenge lies in determining the break-even point compared to traditional LiBs.

### (3) Lithium–sulfur battery

Li–S batteries represent a promising energy storage technology known for their high theoretical energy density $E_{th}$ of 2567 Wh/kg, which can potentially exceed that of conventional LiBs. This high energy density is attributed to the conversion reaction between the lithium metal anode and sulfur cathode (Table 3.1(3)), yielding a cell voltage of 2.2 V.

Additionally, Li–S batteries offer environmental benefits due to the abundance, low cost, and non-toxic nature of sulfur compared to other cathode materials, such as NMC. However, the electrochemical mechanism of Li–S batteries, involving the formation of soluble polysulfides ($Li_2S_n$, where $n = 4$–$8$), presents challenges such as the shuttle effect. This phenomenon causes self-discharge, capacity fading, and the passivation of the lithium–metal surface with insoluble products, thereby further hindering battery performance through the formation of resistive layers on the anode. Furthermore, sulfur undergoes complex compositional and structural changes during charge–discharge cycles, including approximately 80% volume expansion. This significant volume fluctuation, combined with the formation of soluble polysulfide intermediates, contributes to poor mechanical stability of the cathode structure. Lastly, both sulfur and its discharge product, lithium sulfide ($Li_2S$), exhibit electronic and ionic insulating properties. This results in inadequate overall conductivity of the cathode material, leading to low electrochemical utilization of sulfur and rapid capacity fading [7, 8].

The combination of these challenges accounts for the ongoing inadequate long-term stability and cycle performance, which has prevented Li–S batteries from gaining significant commercial traction. Various cathode designs have been explored to address these issues. One such approach, referred to hereafter as "Li–S C excess," involves using a surplus of carbon black as a conductive additive to enhance the electrical conductivity of the sulfur cathode [9]. However, this method primarily relies on the physical confinement of polysulfides, which may not completely prevent the shuttle effect. Moreover, high sulfur loading can still lead to capacity fading over long cycles due to polysulfide dissolution and volume expansion.

Another cathode design, referred to hereafter as "Li–S polymer matrix," incorporates a conductive polymer matrix to enhance cathodic conductivity and mechanical stability [10]. In this approach, conductive polyaniline (PANI) nanotubes are used to encapsulate sulfur, creating a robust framework that accommodates sulfur and enhances electronic conductivity. This polymer matrix improves mechanical stability and flexibility, accommodating volume changes during cycling and enhancing the retention of active material. While this approach shows promise in boosting the overall performance of Li–S batteries, the fabrication process for polymer-based cathodes can be more intricate and costly. Additionally, ensuring long-term stability and compatibility of the polymer with other cell components requires thorough assessment.

The optimization of cathode design and the incorporation of passive components such as conductive and stability additives contribute additional weight to the cell. Consequently, the resulting reduction in energy density is discussed in more detail for the two examples "Li–S C excess" and "Li–S polymer matrix," in Section 3.2.2.

### (4) Lithium–air battery

The principle behind Li–air batteries involves utilizing oxygen from the air as the cathode active material, combined with a lithium metal anode (Figure 3.2, Li–air (n.-aq.)). The resulting discharge reaction equation, presented in Table 3.1(4), yields an exceptionally high theoretical energy density $E_{th}$ of 3505 Wh/kg, positioning Li–air batteries as a promising frontier in energy storage. The use of oxygen as the cathode active material eliminates the need for storing heavy and expensive cathode materials, thereby potentially offering environmental and cost advantages.

However, Li–air batteries face significant challenges related to electrolyte stability and performance, which hinder their practical implementation. Non-aqueous (n.-aq.) organic electrolytes (Figure 3.1, Li–air (n.-aq.)) encounter an issue where the reduction products of the oxygen cathode act as strong reducing agents. These agents decompose carbonate-based solvents, resulting in the formation of insulating carbonate species that cover the electrode surface, rather than forming $Li_2O_2$ reversibly. This process not only reduces reversibility but also compromises cycle performance, emphasizing the necessity for solvent alternatives such as ether-based solvents (e.g., dimethoxyethane ($C_4H_{10}O_2$, DME) and tetraethylene glycol dimethyl ether ($C_{10}H_{22}O_5$, TEGDME)) or aqueous-based electrolytes (such as LiOH or $HClO_3$) dissolved in water. Aqueous electrolytes (Figure 3.2, Li–air (aq.)) participate in the reaction as reactants, thereby enhancing the transferable charge quantity. However, they require a significant amount of water, limiting energy density. By using strong acids such as $HCLO_3$, this effect can be mitigated, enabling energy densities of approximately 968.68 Wh/kg compared to 476.70 Wh/kg achievable with LiOH [11]. Protecting the lithium–metal anode from aqueous electrolytes is essential to prevent lithium oxidation to LiOH. This requires $Li^+$-conducting but electronically insulating membranes to inhibit dendrite formation and enhance battery stability. The same LiOH issue at the anode can arise from excessive moisture in the incoming air, requiring careful management of air humidity levels and a robust electrolyte system design to minimize hydroxide ion migration to the anode [2, 11].

The design of the air cathode is critical to maximizing battery performance. To maintain high capacity, it is essential to optimize the pathways among oxygen, lithium ions, and electrons within the air electrode, ensuring long-lasting reactions. This optimization requires designing high-porosity cathodes that facilitate efficient gas diffusion and promote reaction kinetics. Additionally, introducing catalysts into the air cathode, as highlighted by Mizuno et al. [12], is essential for further enhancing electrochemical reaction kinetics, thereby improving overall battery efficiency.

To illustrate the impact of necessary design optimization measures on energy density, the calculation of the intermediate energy density $E_{inter}$ in Section 3.2.2. will be performed for a Li–air battery utilizing a non-aqueous TEGDME-based electrolyte and a $MnO_2$ catalyst in the cathode.

### (5) Sodium-ion battery

In SiBs, as in LiBs, only the anode and cathode active materials are responsible for electrochemical charge storage through an intercalation-based process (Figure 3.2). SiBs often utilize cathode materials similar to those used in LiBs, such as the LFP analog $Na\,FePO_4$ (NFP) or the NMC analog $NaMn_{0.33}Fe_{0.33}Ni_{0.33}O_2$ (MFN), making their functionality closely resemble that of LiBs. Consequently, SiBs represent a promising "drop-in" technology for LiBs, providing several significant advantages. One of the primary benefits of SiBs is their environmental friendliness. They can utilize eco-friendly materials, such as Prussian white ($Na_2Fe[Fe(CN)_6]$) and Prussian blue ($Fe_4[Fe(CN)_6]$), enhancing their sustainability profile. Additionally, a wide range of active materials available for SiBs allows for greater flexibility in battery design and optimization. SiBs demonstrate outstanding cycling performance, indicating high durability and a long lifespan. They also exhibit high power capabilities, making them suitable for applications requiring quick charging and discharging cycles. Furthermore, SiB can operate effectively over a broader temperature range compared to some lithium-ion variants, increasing their versatility. However, achieving a competitive energy density remains a significant challenge for SiBs. While they are potential substitutes for LFP-based LiBs, they still lag in energy density compared to other LiB chemistries. Another major issue is the larger size of sodium ions compared to lithium ions, which precludes the use of graphite as an anode material. Instead, hard carbon (HC) must be used, complicating the manufacturing process. For an SiB with HC as the anode and NFM as the cathode, the cell voltage results in 3.0 V and the theoretical energy density is calculated to be 274 Wh/kg (Table 3.1(5)), which is approximately 20% less than that of the LiB analog. Despite these challenges, SiBs hold great potential as an alternative to LiBs. Continued research and development are necessary to fully realize their potential and overcome these obstacles, particularly in terms of improving energy density and addressing the requirements for anode materials.

## (6) Zinc battery

Due to the fundamentally different energy storage mechanisms in ZBs, which vary based on the type of electrolyte and cell voltage, using the generalized term "zinc-ion" would be misleading. Zinc-based batteries can be classified into two primary energy storage mechanisms: intercalation-based zinc-ion batteries (ZiBs) and zinc batteries using an aqueous electrolyte that participates in the reaction as a reactant (ZaBs). Figure 3.2 shows a selection of possible active cell components for ZiBs and ZaBs.

ZBs were notably advanced by Xu *et al.* [13], who demonstrated significantly improved rechargeability using a mildly acidic zinc sulfate electrolyte. Their study proposed the intercalation of zinc ions into

$\alpha$-MnO$_2$ as the reaction mechanism, thereby establishing the term "zinc-ion battery" by analogy to intercalation mechanisms used in other battery types. Subsequent research by Lee *et al.* [14] highlighted pH fluctuations during battery cycling, linking MnO$_2$ dissolution and zinc hydroxide sulfate precipitation to these fluctuations. Since then, the storage mechanism at the cathode has been a subject of controversial discussions in the literature. There are even debates about whether intercalation occurs at all, emphasizing the distinction between ZiB and ZaB mechanisms [15].

Further findings by Chao *et al.* [16] identified that different potential ranges are associated with specific reaction mechanisms in ZBs. The upper potential range (1.7–2.0 V) was attributed to electrolytic deposition and dissolution, which can be enhanced by the addition of sulfuric acid (H$_2$SO$_4$). Based on these findings, a 1 M ZnSO$_4$/1 M MnSO$_4$ solution in a mildly acidic aqueous environment is established as the electrolyte for the ZaB system for all subsequent discussions in this chapter. MnO$_2$ is designated as the cathode active material, and the cell voltage is set at 1.5 V, yielding a theoretical energy density $E_{th}$ of 231 Wh/kg for ZaB (Table 3.1(6b)). For enhanced comparability, MnO$_2$ is specified as the cathode active material for the ZiB system as well. With a cell voltage of 1.3 V, the theoretical energy $E_{th}$ of the ZiB results in 291 Wh/kg (Table 3.1(6a)). Given that the precise reaction mechanisms and their interactions are still under debate, the reaction equations provided should be considered simplified representations of the actual processes.

Despite the intricate nature of their reaction mechanisms and low energy density, ZBs remain a widely debated candidate for energy storage, particularly for stationary applications. This is primarily due to their significant material availability and environmental advantages compared to LiBs and SiBs. ZBs can be produced under standard conditions without requiring an inert gas atmosphere, which has the potential to reduce costs [17]. They also provide enhanced safety by utilizing aqueous, non-flammable electrolytes, which minimize thermal fluctuations thanks to the high heat capacity of water.

Nevertheless, challenges persist for zinc-based batteries, including rapid capacity degradation and a shortened lifespan due to hydrogen evolution, corrosion, and dendrite formation. Corrosion-induced hydrogen production contributes to internal pressure buildup and battery dehydration [18]. Addressing these challenges necessitates optimizing anode loadings and efficiency to achieve commercial viability.

### (7) **Lead–acid battery**

Pb batteries remain the leaders in terms of cost in the energy storage market, providing a compelling example of how a seemingly outdated technology has evolved over 160 years to achieve high levels of efficiency. Unlike ion-based battery technologies, charge storage in Pb batteries does not occur through an intercalation process in the electrode active materials; instead, it occurs through a conversion reaction with the aqueous $H_2SO_4$ electrolyte. In Pb batteries, the electrolyte plays an electrochemically active role alongside the Pb anode and the $PbO_2$ cathode (Figure 3.2). The resulting discharge reaction equation is shown in Table 3.1(7), with the cell voltage of the Pb battery at 2.0 V and the theoretical energy density calculated as 167 Wh/kg. Despite not employing the advanced principles of ion-based batteries, Pb batteries have undergone significant technological refinements, primarily driven by their widespread application as starter batteries in vehicles. This evolution underscores how specific applications can propel the development of a technology, as similarly observed with zinc–air batteries in hearing aids.

The advancements in Pb battery technology are notable, with modern Pb batteries now capable of lasting up to 10 years in automotive applications. This extensive development history instills hope for the future potential to surpass the energy density limitations of current ion-based battery technologies.

Furthermore, Pb batteries boast an impressive recycling rate, enhancing their environmental benefits and sustainability credentials. The progress achieved in this long-standing battery technology suggests that with continued innovation, Pb batteries could play a significant role in future energy storage solutions, potentially challenging the dominance of ion-based batteries in terms of energy density.

### 3.2.2 *Calculation of the intermediate energy density $E_{inter}$ incorporating passive cell components*

In the preceding section, the evaluation of energy density across different cell chemistries focused solely on their active cell components that were involved in the reaction equation as reactants (Table 3.1). However, effective functioning of a battery cell requires consideration of various passive components that do not directly participate in the reaction but are crucial nonetheless. These passive cell components add weight to the cell, thereby significantly reducing the achievable energy density compared to the theoretical maximum $E_{th}$.

This section examines the influence of passive cell components on energy density across seven selected cell chemistries, using the exemplary cell compositions detailed in Table 3.2. The passive cell components considered can be categorized as follows:

- Binder in cathode/anode/SSE: $Bind_{CA/AN/SSE}$
- Conductive additive (Cond. add.) in cathode/anode: $C_{CA/AN}$
- Liquid electrolyte fraction (porosity) in cathode/anode/separator: $Liq_{CA/AN/Sep}$
- Solid fraction of the separator: $Sep_{Sep}$
- SSE in separator and cathode: $SSE_{SSE}$ and $SSE_{CA}$
- Excess metallic lithium/zinc: $Li_{excess}$ and $Zn_{excess}$
- Conductive polymer matrix in cathode: $Matx_{CA}$
- Catalyst in cathode: $Cat_{CA}$

Using the cell compositions in Table 3.2 and the reaction equations introduced in the previous section (Table 3.1), a volumetric distribution of the cell components can be derived for each cell chemistry. This results in an overall reaction equation (Table 3.5) that accounts for both active and passive cell components, enabling the calculation of the intermediate energy density $E_{inter}$. The detailed procedure is demonstrated using the example of LiB (Table 3.4). All considered cell components and their relevant material properties are listed in Table 3.3. Many of these material properties vary depending on the manufacturer, necessitating some specifications to be estimated based on literature data and empirical values.

### (1) **Lithium-ion battery**

In LiBs, electrodes are conventionally configured with porous coatings on current collector foils. These coatings include active materials (AM), binders (Bind) to enhance adhesion, and conductive additives (C) to improve electrical conductivity. To prevent direct electrical contact between the anode (AN) and cathode (CA), porous membranes, known as separators (Sep), serve as mechanical barriers. Furthermore, to ensure sufficient ion conduction in the electrodes and in the separator, the pores are filled with a liquid electrolyte (Liq). For the calculation of intermediate energy density $E_{inter}$, the cell composition listed in Table 3.2(1) is assumed, based on the cell specifications given by Chen et al. [19]. The quantity of passive

**Table 3.2.** Specification of use cases for considered cell chemistries, containing the cell composition and the assumed active and passive components used for the calculation of intermediate energy density $E_{inter}$.

| Nr. | System | Cell component | | Material | Weight fraction wt.% | Porosity vol.% | Thickness $\mu$m |
|---|---|---|---|---|---|---|---|
| (1) | LiB | Anode | AN-AM | $C_6$ | 90 | 25 | 25 [1] |
| | | | $C_{AN}$ | C | 5 | | |
| | | | $Bind_{AN}$ | PVDF | 5 | | |
| | | Separator | $Liq_{Sep}$ | 1M LiPF$_6$ in EMC | | 55 | 12 |
| | | | $Sep_{Sep}$ | PP | | | |
| | | Cathode | CA-AM | NMC811 | 90 | 33 | 75.6 |
| | | | $C_{CA}$ | C | 5 | | |
| | | | $Bind_{CA}$ | PVDF | 5 | | |
| (2a) | LiM | Anode | AN-AM | Li (excess) | 100 | 0 | 20 |
| | | Separator | $Liq_{Sep}$ | 1M LiPF$_6$ in EMC | | 55 | 12 |
| | | | $Sep_{Sep}$ | PP | | | |
| | | Cathode | CA-AM | NMC811 | 90 | 33 | 75.6 |
| | | | $C_{CA}$ | C | 5 | | |
| | | | $Bind_{CA}$ | PVDF | 5 | | |
| (2b i.) | Sulfide ASSB | Anode | AN-AM | Li (excess) | 100 | 0 | 20 |
| | | Separator | $SSE_{Sep}$ | LiPSCl | 98 | 0 | 20 |
| | | | $Bind_{Sep}$ | PVDF | 2 | | |
| | | Cathode | CA-AM | NMC811 | 67.2 | 0 | 75.6 |
| | | | $C_{CA}$ | C | 2 | | |
| | | | $Bind_{CA}$ | PVDF | 2 | | |
| | | | $SSE_{CA}$ | LiPSCl | 28.8 | | |
| (2b ii.) | Oxide ASSB | Anode | AN-AM | Li (excess) | 100 | 0 | 20 |
| | | Separator | $SSE_{Sep}$ | LLZO | 98 | 0 | 20 |
| | | | $Bind_{Sep}$ | PVDF | 2 | | |
| | | Cathode | CA-AM | NMC811 | 67.2 | 0 | 75.6 |
| | | | $C_{CA}$ | C | 2 | | |
| | | | $Bind_{CA}$ | PVDF | 2 | | |
| | | | $SSE_{CA}$ | LLZO | 28.8 | | |
| (3a) | Li-S C excess | Anode | AN-AM | Li (excess) | 100 | 0 | 20 |
| | | Separator | $Liq_{Sep}$ | 1M LiCF$_3$SO$_3$ in TEGDME | | 55 | 25 |
| | | | $Sep_{Sep}$ | PP | | | |
| | | Cathode | CA-AM | S | 56.7 | 30 | 60 |
| | | | $C_{CA}$ | C | 27.3 | | |
| | | | $Bind_{CA}$ | PVDF | 16 | | |

*(Continued)*

**Table 3.2.** *(Continued)*

| Nr. | System | Cell component | | Material | Weight fraction wt.% | Porosity vol.% | Thickness $\mu$m |
|---|---|---|---|---|---|---|---|
| **(3b)** | **Li-S polymer matrix** | **Anode** | AN-AM | Li (excess) | 100 | | 20 |
| | | **Separator** | $Liq_{Sep}$ | 1M LiFSI in DME | | 55 | 12 |
| | | | $Sep_{Sep}$ | PP | | | |
| | | **Cathode** | CA-AM | S | 49.6 | 30 | 60 |
| | | | $C_{CA}$ | C | 10 | | |
| | | | $Bind_{CA}$ | PVDF | 10 | | |
| | | | $Matx_{CA}$ | PANI | 30.4 | | |
| **(4)** | **Li-air** | **Anode** | AN-AM | Li (excess) | 100 | | 20 |
| | | **Separator** | $Liq_{Sep}$ | 1M $LiPF_6$ in TEGDME | | 55 | 12 |
| | | | $Sep_{Sep}$ | PP | | | |
| | | **Cathode** | CA-AM | $Li_2O_2$ | $70^3$ | $20^2$ | 60 |
| | | | $C_{CA}$ | C | 25 ($7.5^3$) | | |
| | | | $Bind_{CA}$ | PTFE | 33 ($9.9^3$) | | |
| | | | $CAT_{CA}$ | $MnO_2$ | 42 ($12.6^3$) | | |
| **(5)** | **SiB** | **Anode** | AN-AM | HC | 90 | 0.3 | $80.21^1$ |
| | | | $C_{AN}$ | C | 5 | | |
| | | | $Bind_{AN}$ | PVDF | 5 | | |
| | | **Separator** | $Liq_{Sep}$ | 1M $NaPF_6$ in EMC | | 0.55 | 12 |
| | | | $Sep_{Sep}$ | PP | | | |
| | | **Cathode** | CA-AM | MFN | 90 | 0.3 | 55 |
| | | | $C_{CA}$ | C | 5 | | |
| | | | $Bind_{CA}$ | PVDF | 5 | | |
| **(6a,b)** | **ZB** | **Anode** | AN-AM | Zn (excess) | 100 | | ZiB: $40^4$ ZaB: $80^4$ |
| | | **Separator** | $Liq_{Sep}$ | ZiB: 2M $ZnSO_4$ in $H_2O$ ZaB: 1M $ZnSO_4$ + 1M $MnSO_4$ in $H_2O$ | | 0.55 | 15 |
| | | | $Sep_{Sep}$ | PP | | | |
| | | **Cathode** | CA-AM | $MnO_2$ | 96 | 0.25 | 98 |
| | | | $C_{CA}$ | C | 2 | | |
| | | | $Bind_{CA}$ | PTFE | 2 | | |

**Table 3.2.** (*Continued*)

| Nr. | System | Cell component | | Material | Weight fraction wt.% | Porosity vol.% | Thickness $\mu$m |
|---|---|---|---|---|---|---|---|
| (7) | Pb | Anode | AN-AM | Pb | 96 | 0.53 | 703[1] |
| | | | $C_{AN}$ | C | 2 | | |
| | | | $Bind_{AN}$ | LS | 2 | | |
| | | Separator | $Liq_{Sep}$ | 40wt.% $H_2SO_4$ in $H_2O$ | | 0.92 | 1500 |
| | | | $Sep_{Sep}$ | PE | | | |
| | | Cathode | CA-AM | $PbO_2$ | 96 | 0.57 | 1250 |
| | | | $C_{CA}$ | C | 2 | | |
| | | | $Bind_{CA}$ | LS | 2 | | |

*Note*: [1]The anode thickness $L_{AN}$ is calculated based on the reaction stoichiometry and parameters above.
[2]Assumed porosity after the discharge reaction [2].
[3]The $Li_2O_2$ mass fraction is fixed to 70 wt.%, which is approximately 60 vol.% [2]. Based on the recipe of Mizuno *et al.* [12], the mass fractions of the passive cathode components are 25:33:42 wt.%. Taking into account the $Li_2O_2$ proportion of the total mass, these result in 7.5:9.9:12.6 wt.%.
[4]Calculated Zn thickness based on an assumed N/P ratio of 1.1 [4].

**Table 3.3.** Cell material properties.

| Material | Abbreviation | Molar mass (g/mol) | Density (g/cm³) | Refs. |
|---|---|---|---|---|
| | **Cathode** | | | |
| $LiNi_{0.8}Co_{0.1}Mn_{0.1}O_2$ | NMC811 | 97.28 | 2.2 | [20] |
| $Li_{0.5}Ni_{0.8}Co_{0.1}Mn_{0.1}O_2$ | NMC811 | 93.81 | 2.2 | [20] |
| $Li_2S$ | — | 45.95 | 1.67 | [21] |
| S | | 32.06 | 2.07 | |
| $Li_2O_2$ | — | 45.88 | 2.33 | [22] |
| $O_2$ | — | 32.00 | 0.00014 | |
| $NaMn_{0.33}Fe_{0.33}Ni_{0.33}O_2$ | MFN | 110.92 | 5 | Estimation |
| $Na_{0.5}Mn_{0.33}Fe_{0.33}Ni_{0.33}O_2$ | MFN | 99.42 | 5 | Estimation |
| $MnO_2$ | — | 86.94 | 5.03 | [23] |
| $MnSO_4$ | — | 151 | 3.49 | [24] |
| $PbSO_4$ | — | 303.256 | 6.34 | [25] |
| $PbO_2$ | — | 239.2 | 9.34 | [26] |

(*Continued*)

58 *K. Schad & K. P. Birke*

**Table 3.3.** (*Continued*)

| Material | Abbreviation | Molar mass (g/mol) | Density (g/cm³) | Refs. |
|---|---|---|---|---|
| | **Anode** | | | |
| $LiC_6$ | — | 79.01 | 2.16 | Estimation |
| $C_6$ | — | 72.07 | 2.2 | [27] |
| Li | — | 6.94 | 0.53 | |
| $NaC_6$ | — | 95.06 | 2.01 | Estimation |
| $ZnSO_4$ | — | 161.47 | 3.93 | [28] |
| Zn | — | 65.38 | 7.14 | |
| Pb | — | 207.2 | 11.34 | |
| | **Liquid/solid-state electrolyte** | | | |
| 1M $LiPF_6$ in $C_4H_6O_3$ | 1M $LiPF_6$ in EMC | 151.91 | 1.21 | Estimation |
| 1M $F_2LiNO_4S_2$ in $C_3H_3FO_3$ | 1M LiFSI in FEC | 187.05 | 1.35 | Estimation |
| $Li_6PS_5Cl$ | LiPSCl | 268.398 | 1.64 | [29] |
| $Li_7La_3Zr_2O_{12}$ | LLZO | 839.753 | 5.01 | [30] |
| 1M $LiPF_6$ in $C_{10}H_{22}O_5$ | 1M $LiPF_6$ in TEGDME | 151.91 | 1.13 | Estimation |
| 1M $LiCF_3SO_3$ in $C_{10}H_{22}O_5$ | 1M LiOTf in TEGDME | 156.01 | 1.07 | Estimation |
| 1M $F_2LiNO_4S_2$ in $C_4H_{10}O_2$ | 1M LiFSI in DME | 187.05 | 1.07 | Estimation |
| 1M $NaPF_6$ in $C_4H_6O_3$ | 1M $NaPF_6$ in EMC | 167.95 | 1.12 | Estimation |
| 1M $ZnSO_4$/1M $MnSO_4$ in $H_2O$ | — | 312.38 | 1.24 | Estimation |
| 2M $ZnSO_4$ in $H_2O$ | — | 161.47 | 1.28 | [31] |
| $H_2O$ | — | 18.015 | 0.998 | |
| $H_2SO_4$ | — | 98.08 | 1.84 | [32] |
| 40% $H_2SO_4$ in $H_2O$ | — | 98.08 | 1.3 | Estimation |
| | **Separator** | | | |
| $[C_3H_6]_n$ | PP | 50000[1] | 0.91 | Estimation |
| $[C_2H_4]_n$ | PE | 4000000[1] | 0.93 | Estimation |
| | **Binder** | | | |
| $[C_2H_2F_2]_n$ | PVDF | 534000[1] | 1.75 | Estimation |
| $[C_8H_{15}NaO_8]_n$ | Na-cmc | 100000[1] | 0.7 | Estimation |
| $[C_2F_4]_n$ | PTFE | 50000[1] | 2.1 | Estimation |
| $[C_{20}H_{24}Na_2O_{10}S_2]_n$ | LS | 100000[1] | 0.7 | Estimation |

**Table 3.3.** (*Continued*)

| Material | Abbreviation | Molar mass (g/mol) | Density (g/cm³) | Refs. |
|---|---|---|---|---|
| | **Conductive additive** | | | |
| C | — | 12.01 | 1.9 | [27] |
| $[C_6H_5N]_n$ | PANI | 3000[1] | 1.35 | estimation |
| | **Catalyst** | | | |
| $MnO_2$ | — | 86.94 | 5.03 | [23] |
| Au | — | 196.97 | 19.32 | |
| Pt | — | 195.08 | 21.45 | |

*Note*: [1]Estimated value for the average molar mass of the polymer.

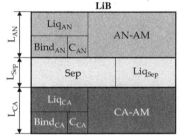

CA-AM: Cathode active material = 1 volume element
AN-AM: Anode active material relatively to CA-AM
Sep: Solid separator material relatively to CA-AM
$Liq_{AN}$: Liquid electrolyte in anode relatively to CA-AM
$Liq_{CA}$: Liquid electrolyte in cathode relatively to CA-AM
$Liq_{Sep}$: Liquid electrolyte in separator relatively to CA-AM
$Bind_{AN}$: Binder in anode relatively to CA-AM
$Bind_{CA}$: Binder in cathode relatively to AN-AM
$C_{AN}$: Conductive additive in anode relatively to CA-AM
$C_{CA}$: Conductive additive in cathode relatively to CA-AM
L: Layer thickness

**Figure 3.3.** Volumetric distribution of the cell components in an LiB.

components within the cell is defined by the mass or volume fractions added during cell production, typically expressed as weight percentage (wt.%) or volume percentage (vol.%).

Translating these recipe specifications into quantities of substances that can be integrated on both sides of the overall reaction equation involves a crucial methodology. A key step here is converting the material quantities utilized during cell production into a volume distribution normalized by the volume of the cathode active material (CA-AM). The resulting volume distribution is schematically represented in Figure 3.3.

The step-by-step calculation of this volume distribution and its translation into the intermediate energy density $E_{inter}$ is detailed in Table 3.4. It begins with the discharge reaction equation, which includes the active cell components. The stoichiometric coefficients represent the amount of substance $n$ in mol of each reactant. Using the known molar

**Table 3.4.**   Example calculation of intermediate energy density $E_{inter}$ for LiB, including equations (Equ.) used.

| | | | Discharge reaction | Equ. |
|---|---|---|---|---|
| **Active components** | $n$ | $n_1$ | $n_1$ LiC$_6$ + $n_2$ Li$_{0.5}$Ni$_{0.8}$Co$_{0.1}$Mn$_{0.1}$O$_2$ − $n_3$ C$_6$ + $n_4$ LiNi$_{0.8}$Co$_{0.1}$Mn$_{0.1}$O$_2$ | |
| | mol | 1 | 1 LiC$_6$ + 2 Li$_{0.5}$Ni$_{0.8}$Co$_{0.1}$Mn$_{0.1}$O$_2$ − 1 C$_6$ + 2 LiNi$_{0.8}$Co$_{0.1}$Mn$_{0.1}$O$_2$ | |
| | $m$ | $m_1$ | $m_1$ LiC$_6$ + $m_2$ Li$_{0.5}$Ni$_{0.8}$Co$_{0.1}$Mn$_{0.1}$O$_2$ − $m_3$ C$_6$ + $m_4$ LiNi$_{0.8}$Co$_{0.1}$Mn$_{0.1}$O$_2$ | (2) |
| | g | 79.0 | 79.0 LiC$_6$ + 187.6 Li$_{0.5}$Ni$_{0.8}$Co$_{0.1}$Mn$_{0.1}$O$_2$ − 72.1 C$_6$ + 194.6 LiNi$_{0.8}$Co$_{0.1}$Mn$_{0.1}$O$_2$ | |
| | $V$ | $V_1$ | $V_1$ LiC$_6$ + $V_2$ Li$_{0.5}$Ni$_{0.8}$Co$_{0.1}$Mn$_{0.1}$O$_2$ − $V_3$ C$_6$ + $V_4$ LiNi$_{0.8}$Co$_{0.1}$Mn$_{0.1}$O$_2$ | (3) |
| | cm$^3$ | 36.6 | 36.6 LiC$_6$ + 85.3 Li$_{0.5}$Ni$_{0.8}$Co$_{0.1}$Mn$_{0.1}$O$_2$ − 32.76 C$_6$ + 88.4 LiNi$_{0.8}$Co$_{0.1}$Mn$_{0.1}$O$_2$ | |
| | $V_n$ | $V_{n,1}$ | $V_{n,1}$ LiC$_6$ + $V_{n,2}$ Li$_{0.5}$Ni$_{0.8}$Co$_{0.1}$Mn$_{0.1}$O$_2$ − $V_{n,3}$ C$_6$ + $V_{n,4}$ LiNi$_{0.8}$Co$_{0.1}$Mn$_{0.1}$O$_2$ | (4) |
| | - | 0.41 | 0.41 LiC$_6$ + 0.96 Li$_{0.5}$Ni$_{0.8}$Co$_{0.1}$Mn$_{0.1}$O$_2$ − 0.37 C$_6$ + 1.00 LiNi$_{0.8}$Co$_{0.1}$Mn$_{0.1}$O$_2$ | |
| | $E_{th}$ | | **372 Wh/kg** | (1) |
| **Active + passive components** | $V_n$ | | $V_{n,1}$ LiC$_6$ + $V_{n,2}$ Li$_{0.5}$Ni$_{0.8}$Co$_{0.1}$Mn$_{0.1}$O$_2$ + $V_{n,P1}$ Liq$_{AN}$ + $V_{n,P2}$ Bind$_{AN}$ + $V_{n,P3}$ C$_{AN}$ + $V_{n,P4}$ Sep$_{Sep}$ + | (4) |
| | - | | $V_{n,P5}$ Liq$_{Sep}$ + $V_{n,P6}$ Liq$_{CA}$ + $V_{n,P7}$ Bind$_{CA}$ + $V_{n,P8}$ C$_{CA}$ | |
| | | → | $V_{n,3}$ C$_6$ + $V_{n,4}$ LiNi$_{0.8}$Co$_{0.1}$Mn$_{0.1}$O$_2$ + $V_{n,P1}$ Liq$_{AN}$ + $V_{n,P2}$ Bind$_{AN}$ + $V_{n,P3}$ C$_{AN}$ + $V_{n,P4}$ Sep$_{Sep}$ + | |
| | | | $V_{n,P5}$ Liq$_{Sep}$ + $V_{n,P6}$ Liq$_{CA}$ + $V_{n,P7}$ Bind$_{CA}$ + $V_{n,P8}$ C$_{CA}$ | |
| | | | 0.41 LiC$_6$ + 0.96 Li$_{0.5}$Ni$_{0.8}$Co$_{0.1}$Mn$_{0.1}$O$_2$ + 0.14 Liq$_{AN}$ + 0.03 Bind$_{AN}$ + 0.02 C$_{AN}$ + 0.12 Sep$_{Sep}$ + | |
| | | | 0.15 Liq$_{Sep}$ + 0.56 Liq$_{CA}$ + 0.07 Bind$_{CA}$ + 0.06 C$_{CA}$ | |
| | | → | 0.37 C$_6$ + 1.00 LiNi$_{0.8}$Co$_{0.1}$Mn$_{0.1}$O$_2$ + 0.14 Liq$_{AN}$ + 0.03 Bind$_{AN}$ + 0.02 C$_{AN}$ + 0.12 Sep$_{Sep}$ + | |
| | | | 0.15 Liq$_{Sep}$ + 0.56 Liq$_{CA}$ + 0.07 Bind$_{CA}$ + 0.06 C$_{CA}$ | |

$$
\begin{aligned}
f \quad & f_1 \quad \text{LiC}_6 + f_2 \quad \text{Li}_{0.5}\text{Ni}_{0.8}\text{Co}_{0.1}\text{Mn}_{0.1}\text{O}_2 + f_{P1} \quad \text{Liq}_{AN} + f_{P2} \quad \text{Bind}_{AN} + f_{P3} \quad \text{C}_{AN} + f_{P4} \quad \text{Sep}_{Sep} + (2), \\
\text{mol} \quad & f_{P5} \quad \text{Liq}_{Sep} + f_{P6} \quad \text{Liq}_{CA} \qquad\qquad + f_{P7} \quad \text{Bind}_{CA} + f_{P8} \quad \text{C}_{CA} \hfill (3) \\
\rightarrow & f_3 \quad \text{C}_6 + f_4 \quad \text{LiNi}_{0.8}\text{Co}_{0.1}\text{Mn}_{0.1}\text{O}_2 + f_{P1} \quad \text{Liq}_{AN} + f_{P2} \quad \text{Bind}_{AN} + f_{P3} \quad \text{C}_{AN} + f_{P4} \quad \text{Sep}_{Sep} + \\
& f_{P5} \quad \text{Liq}_{Sep} + f_{P6} \quad \text{Liq}_{CA} \qquad\qquad + f_{P7} \quad \text{Bind}_{CA} + f_{P8} \quad \text{C}_{CA} \\
& \mathbf{0.01} \quad \text{LiC}_6 + \mathbf{0.02} \quad \text{Li}_{0.5}\text{Ni}_{0.8}\text{Co}_{0.1}\text{Mn}_{0.1}\text{O}_2 + \mathbf{1E{-}03} \quad \text{Liq}_{AN} + \mathbf{9E{-}08} \quad \text{Bind}_{AN} + \mathbf{0.004} \quad \text{C}_{AN} + \mathbf{4E{-}06} \quad \text{Sep}_{Sep} + \\
& \mathbf{1E{-}3} \quad \text{Liq}_{Sep} + \mathbf{4E{-}3} \quad \text{Liq}_{CA} \qquad\qquad + \mathbf{2E{-}07} \quad \text{Bind}_{CA} + \qquad \text{C}_{CA} \\
\rightarrow & \mathbf{0.01} \quad \text{C}_6 + \mathbf{0.02} \quad \text{LiNi}_{0.8}\text{Co}_{0.1}\text{Mn}_{0.1}\text{O}_2 + \mathbf{1E{-}03} \quad \text{Liq}_{AN} + \mathbf{9E{-}08} \quad \text{Bind}_{AN} + \mathbf{0.004} \quad \text{C}_{AN} + \mathbf{4E{-}06} \quad \text{Sep}_{Sep} + \\
& \mathbf{1E{-}3} \quad \text{Liq}_{Sep} + \mathbf{4E{-}3} \quad \text{Liq}_{CA} \qquad\qquad + \mathbf{2E{-}07} \quad \text{Bind}_{CA} + \mathbf{0.01} \quad \text{C}_{CA}
\end{aligned}
$$

$$
E_{inter} \qquad\qquad\qquad\qquad\qquad\qquad \mathbf{250 \ Wh/kg} \qquad\qquad\qquad\qquad\qquad\qquad (1)
$$

mass $M$ in g/mol and density $\rho$ in g/cm³, the amounts of substances of the active cell components can be converted into volumes $V$ in cm³ using equations (3.2) and (3.3):

$$m = Mn, \tag{3.2}$$

$$V = \frac{m}{\rho}. \tag{3.3}$$

$V_n$, the volume relative to the cathode active material (CA-AM), is obtained by normalizing the individual volumes with $V_4$, the volume of the CA-AM in cm³ (equation (3.4)):

$$V_n = \frac{1}{V_4}. \tag{3.4}$$

The mass fractions specified in the electrode recipe (Table 3.2) establish the mathematical relationship between the solid passive electrode components ($Bind_{CA/AN}$ and $C_{CA/AN}$), the anode active material (AN-AM), and CA-AM. Using density, these mass fractions can also be converted into volumes and normalized to the respective volumes of AN-AM and CA-AM. The contribution of the electrolyte in the electrodes ($Liq_{CA/AN}$) is determined by the given porosity in vol.% and can similarly be normalized. For the cathode, the volume fractions of the passive cell components relative to CA-AM are denoted as $V_{n,P6-8}$. Given the fixed stoichiometric ratio between AN-AM and CA-AM, as defined by the reaction equation, the normalized volumes of the passive anode components are multiplied by $V_{n,1}$, representing the volume fraction of the anode relative to CA-AM. This calculation yields $V_{n,P1-3}$, which denote the volume fractions of the passive anode materials relative to CA-AM. The relationship between the separator components, $Liq_{Sep}$ and $Sep_{Sep}$, and CA-AM is defined by the cathode thickness $L_{Ca}$ and the separator thickness $L_{Sep}$. The effective CA-AM thickness $L_{CA-AM}$ is derived from the volume fraction of CA-AM in the total cathode volume and is subsequently related to $L_{Sep}$. Given the separator porosity, the volume fractions of $Liq_{Sep}$ and $Sep_{Sep}$ relative to CA-AM are determined as $V_{n,P4-5}$. Using these calculated volume fractions relative to CA-AM ($V_{n,P}$) as coefficients, the passive cell components can now be integrated on both sides of the reaction equation. Using the molar mass and density (equations (3.2) and (3.3)), the amounts of substance are recalculated as $f$. The index $P$ is employed to clearly distinguish the amounts of substance of the passive components from those of the active components. The terms in the overall reaction equation, obtained by weighting the passive components with these factors $f_p$, are referred to as BIRKE-summands. This yields the complete discharge reaction equation and the intermediate energy density $E_{inter} = 250$ Wh/kg using equation (3.1). In this LiB use case, the passive cell components reduce the theoretical energy density $E_{th}$ by 33%. An overview of the complete discharge reaction equations for all considered cell chemistries is provided in Table 3.5. Detailed discussions of individual equations are provided in the corresponding section on each cell chemistry.

**Table 3.5.** Overall discharge reaction with BIRKE-summands for calculating intermediate energy density $E_{inter}$.

| Nr. | System | Discharge reaction |
|-----|--------|--------------------|
| (1) | LiB | $f_1 \text{LiC}_6 + f_2 \text{Li}_{0.5}\text{Ni}_{0.8}\text{Co}_{0.1}\text{Mn}_{0.1}\text{O}_2 + f_{P1} \text{Liq}_{AN} + f_{P2} \text{Bind}_{AN} + f_{P3} \text{C}_{AN} + f_{P4} \text{Sep}_{Sep} + f_{P5} \text{Liq}_{Sep} + f_{P6} \text{Liq}_{Ca} + f_{P7} \text{Bind}_{CA} + f_{P8} \text{C}_{CA}$ <br> $\longrightarrow$ <br> $f_3 \text{C}_6 + f_4 \text{LiNi}_{0.8}\text{Co}_{0.1}\text{Mn}_{0.1}\text{O}_2 + f_{P1} \text{Liq}_{AN} + f_{P2} \text{Bind}_{AN} + f_{P3} \text{C}_{AN} + f_{P4} \text{Sep}_{Sep} + f_{P5} \text{Liq}_{Sep} + f_{P6} \text{Liq}_{Ca} + f_{P7} \text{Bind}_{CA} + f_{P8} \text{C}_{CA}$ |
| (2a) | LiM | $f_1 \text{Li} + f_2 \text{Li}_{0.5}\text{Ni}_{0.8}\text{Co}_{0.1}\text{Mn}_{0.1}\text{O}_2 + f_{P4} \text{Sep}_{Sep} + f_{P5} \text{Liq}_{Sep} + f_{P6} \text{Liq}_{Ca} + f_{P7} \text{Bind}_{CA} + f_{P8} \text{C}_{CA} + f_{P9} \text{Li}_{excess}$ <br> $\longrightarrow$ <br> $f_4 \text{LiNi}_{0.8}\text{Co}_{0.1}\text{Mn}_{0.1}\text{O}_2 + f_{P4} \text{Sep}_{Sep} + f_{P5} \text{Liq}_{Sep} + f_{P6} \text{Liq}_{Ca} + f_{P7} \text{Bind}_{CA} + f_{P8} \text{C}_{CA} + f_{P9} \text{Li}_{excess}$ |
| (2b) | ASSB | $f_1 \text{Li} + f_2 \text{Li}_{0.5}\text{Ni}_{0.8}\text{Co}_{0.1}\text{Mn}_{0.1}\text{O}_2 + f_{P4} \text{Bind}_{SSE} + f_{P5} \text{SSE}_{SSE} + f_{P6} \text{SSE}_{CA} + f_{P7} \text{Bind}_{CA} + f_{P8} \text{C}_{CA} + f_{P9} \text{Li}_{excess}$ <br> $\longrightarrow$ <br> $f_4 \text{LiNi}_{0.8}\text{Co}_{0.1}\text{Mn}_{0.1}\text{O}_2 + f_{P4} \text{Bind}_{SSE} + f_{P5} \text{SSE}_{SSE} + f_{P6} \text{SSE}_{CA} + f_{P7} \text{Bind}_{CA} + f_{P8} \text{C}_{CA} + f_{P9} \text{Li}_{excess}$ |
| (3a) | Li–S C excess | $f_1 \text{Li} + f_2 \text{S} + f_{P4} \text{Sep}_{Sep} + f_{P5} \text{Liq}_{Sep} + f_{P6} \text{Liq}_{Ca} + f_{P7} \text{Bind}_{CA} + f_{P8} \text{C}_{CA} + f_{P9} \text{Li}_{excess}$ <br> $\longrightarrow$ <br> $f_4 \text{Li}_2\text{S} + f_{P4} \text{Sep}_{Sep} + f_{P5} \text{Liq}_{Sep} + f_{P6} \text{Liq}_{Ca} + f_{P7} \text{Bind}_{CA} + f_{P8} \text{C}_{CA} + f_{P9} \text{Li}_{excess}$ |
| (3b) | Li–S polymer matrix | $f_1 \text{Li} + f_2 \text{S} + f_{P4} \text{Sep}_{Sep} + f_{P5} \text{Liq}_{Sep} + f_{P6} \text{Liq}_{Ca} + f_{P7} \text{Bind}_{CA} + f_{P8} \text{C}_{CA} + f_{P9} \text{Li}_{excess} + f_{P10} \text{Matx}_{CA}$ <br> $\longrightarrow$ <br> $f_4 \text{Li}_2\text{S} + f_{P4} \text{Sep}_{Sep} + f_{P5} \text{Liq}_{Sep} + f_{P6} \text{Liq}_{Ca} + f_{P7} \text{Bind}_{CA} + f_{P8} \text{C}_{CA} + f_{P9} \text{Li}_{excess} + f_{P10} \text{Matx}_{CA}$ |

*(Continued)*

## Table 3.5. (*Continued*)

| Nr. | System | Discharge reaction |
|---|---|---|
| (4) | Li–air | $f_1\ Li + f_2\ O_2 + f_{P4}\ Sep_{Sep} + f_{P5}\ Liq_{Sep} + f_{P6}\ Liq_{Ca} + f_{P7}\ Bind_{CA} + f_{P8}\ C_{CA} + f_{P9}\ Li_{excess} + f_{P10}\ CAT_{CA}$ $\rightarrow$ $f_4\ Li_2O_2 + f_{P4}\ Sep_{Sep} + f_{P5}\ Liq_{Sep} + f_{P6}\ Liq_{Ca} + f_{P7}\ Bind_{CA} + f_{P8}\ C_{CA} + f_{P9}\ Li_{excess} + f_{P10}\ CAT_{CA}$ |
| (5) | SiB | $f_1\ NaC_6 + f_2\ Na_{0.5}Ni_{0.33}Fe_{0.33}Mn_{0.33}O_2 + f_{P1}\ Liq_{AN} + f_{P2}\ Bind_{AN} + f_{P3}\ C_{AN} + f_{P4}\ Sep_{Sep} + f_{P5}\ Liq_{Sep} + f_{P6}\ Liq_{Ca} + f_{P7}\ Bind_{CA} + f_{P8}\ C_{CA}$ $\rightarrow$ $f_3\ C_6 + f_4\ NaNi_{0.33}Fe_{0.33}Mn_{0.33}O_2 + f_{P1}\ Liq_{AN} + f_{P2}\ Bind_{AN} + f_{P3}\ C_{AN} + f_{P4}\ Sep_{Sep} + f_{P5}\ Liq_{Sep} + f_{P6}\ Liq_{Ca} + f_{P7}\ Bind_{CA} + f_{P8}\ C_{CA}$ |
| (6a) | ZiB | $f_1\ Zn + f_2\ MnO_2 + f_{P4}\ Sep_{Sep} + f_{P5}\ Liq_{Sep} + f_{P6}\ Liq_{Ca} + f_{P7}\ Bind_{CA} + f_{P8}\ C_{CA} + f_{P9}\ Zn_{excess}$ $\rightarrow$ $f_4\ ZnMn_2O_4 + f_{P4}\ Sep_{Sep} + f_{P5}\ Liq_{Sep} + f_{P6}\ Liq_{Ca} + f_{P7}\ Bind_{CA} + f_{P8}\ C_{CA} + f_{P9}\ Zn_{excess}$ |
| (6b) | ZaB | $f_1\ Zn + f_2\ MnO_2 + f_5\ H_2SO_4 + f_{P4}\ Sep_{Sep} + f_{P5}\ Liq_{Sep} + f_{P6}\ Liq_{Ca} + f_{P7}\ Bind_{CA} + f_{P8}\ C_{CA} + f_{P9}\ Zn_{excess}$ $\rightarrow$ $f_3\ ZnSO_4 + f_4\ MnSO_4 + f_6\ H_2O + f_{P4}\ Sep_{Sep} + f_{P5}\ Liq_{Sep} + f_{P6}\ Liq_{Ca} + f_{P7}\ Bind_{CA} + f_{P8}\ C_{CA} + f_{P9}\ Zn_{excess}$ |
| (7) | Pb | $f_1\ Pb + f_2\ PbO_2 + f_5\ H_2SO_4 + f_{P1}\ Liq_{AN} + f_{P2}\ Bind_{AN} + f_{P3}\ C_{AN} + f_{P4}\ Sep_{Sep} + f_{P5}\ Liq_{Sep} + f_{P6}\ Liq_{Ca} + f_{P7}\ Bind_{CA} + f_{P8}\ C_{CA}$ $\rightarrow$ $f_3\ PbSO_4 + f_6\ H_2O + f_{P1}\ Liq_{AN} + f_{P2}\ Bind_{AN} + f_{P3}\ C_{AN} + f_{P4}\ Sep_{Sep} + f_{P5}\ Liq_{Sep} + f_{P6}\ Liq_{Ca} + f_{P7}\ Bind_{CA} + f_{P8}\ C_{CA}$ |

The calculated intermediate energy density $E_{inter}$ of the LiB, however, applies exclusively to the assumed, non-standardized cell composition.

To assess the impact of varying cell compositions on $E_{inter}$, a sensitivity analysis was conducted, considering the following: (a) the mass fractions (or volume fractions for the liquid electrolyte) of the passive components, (b) the thicknesses of the anode, cathode, and separator layers, and (c) the density of the passive components. The following assumptions were made for the parameter variation: The layer thicknesses $L_{AN}$ and $L_{CA}$ and thus the total volume $V_{ges}$ of the electrode affected by the variation are constant. The total mass $m_{solid}$ of the solid components of the electrode affected by the variation is also constant. This implies that adjusting the mass fraction of one solid electrode component necessitates proportional adjustments in the mass fractions of the other electrode components to maintain $m_{solid}$. The resulting change in the solid volume $V_{solid}$ is compensated by an equivalent adjustment in the porosity, which matches the liquid volume $V_{liq}$ of the electrode affected by the variation to maintain $V_{ges}$. In this LiB use case, a weight ratio of 90:5:5 wt.% between CA-AM, $Bind_{CA}$, and $C_{CA}$ is assumed, with 33 vol.% of the total cathode volume $V_{ges,}$ filled with liquid electrolyte. If, for instance, the binder content is increased to 10 wt.%, the weight fractions of CA-AM and $C_{CA}$ are reduced to 85.26 wt.% and 4.73 wt.%, respectively, based on their proportional relationship. Given that the binder density (PVDF,1.75 $g/cm^3$) is lower than the density of CA-AM (NMC, 2.2 $g/cm^3$), this results in an increase in $V_{solid}$. To maintain $V_{ges}$, the cathode porosity $Liq_{CA}$ decreases to 32 vol.%. When varying the volume fraction of the electrode porosity $Liq_{CA}$, changes in the volume fraction of the solid electrode components, and hence $V_{solid}$, occur while maintaining $m_{solid}$. For instance, increasing $Liq_{CA}$ from 33 vol.% to 70 vol.% reduces the solid content from 67 vol.% to 30 vol.%, simultaneously decreasing the volume of CA-AM. Since $m_{solid}$ remains unchanged, the mass of CA-AM within the cell also remains constant, while the liquid mass $m_{liq}$ increases, thereby increasing the total mass of the cathode.

The relationships described above and their impact on the energy density are illustrated in Figure 3.4(a). Anode components are represented in black, cathode components in dark gray, and separator components in light gray. Dotted horizontal lines indicate the theoretical energy density $E_{th}$ and the intermediate energy density $E_{inter}$ of the selected cell use case, as detailed in Table 3.2(1).

The findings indicate that increasing the volume fraction of a passive cell component consistently decreases the energy density. This trend arises because a higher volume fraction of a passive cell component relative to CA-AM increases the mass fraction of passive components in the overall weight, thereby decreasing the mass fraction of the active component and subsequently reducing the energy density. The gradient of the energy density curves varies depending on the density of the components being adjusted and their initial weight ratio. This phenomenon is due to the varying material densities and the initial relation between the passive materials and CA-AM. When starting with an initial ratio of 90:5:5 wt.% between CA-AM, $C_{CA}$, and $Bind_{CA}$, increasing $Bind_{CA}$ by 5 wt.% results in

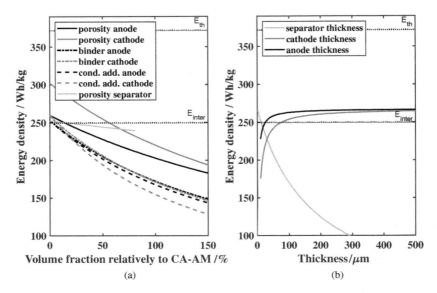

**Figure 3.4.** Energy density curves of an LiB depending on: (a) the volume fraction of cell components relative to CA-AM, and (b) the layer thickness of separator, cathode, and anode. Dotted horizontal lines indicate the theoretical energy density $E_{th}$ and the intermediate energy density $E_{inter}$ for the cell configuration specified in Table 3.2(1).

a more significant reduction in energy density that is more pronounced compared to starting with a ratio of 96:2:2 wt.%, where the mass fraction of CA-AM relative to the other non-varied passive components is higher. When comparing the curves of the initially equally weighted solid anode components $Bind_{AN}$ and $C_{AN}$, they can be ranked according to their respective density values: $Bind_{AN}$ (PVDF, 1.75 g/cm$^3$) < $C_{AN}$ (C, 1.9 g/cm$^3$). In this scenario, where the initial ratios between CA-AM and $C_{CA}$ and between CA-AM and $Bind_{CA}$ are identical, the gradient of the energy density curve is solely determined by density.

So far, the cathode thickness $L_{CA}$ has been assumed to remain constant at 75.6 μm. Based on this assumption and the established stoichiometric relationships between the anode and cathode, the resulting anode thickness $L_{AN}$ is 25 μm (assumed N/P ratio = 1). This relatively low value can be attributed to the assumed delithiation depth of only 50% for the NMC cathode (refer to the reaction equation in Table 3.1).

In Figure 3.4(b), the influence of anode, cathode, and separator thicknesses on the energy density is illustrated. Increasing electrode thickness results in higher energy density due to the increased fraction of active material relative to that of the passive separator material. Reducing separator thickness produces the same effect. To evaluate the impact of passive cell component density, Figure 3.5(a) depicts energy density as a function of material density in g/cm$^3$. The curves result from density variations of passive components while maintaining constant volumes and volume ratios relative to CA-AM. These density variations alter the mass contributions of individual components, directly affecting energy density. To provide additional context, specific density variations for the electrolyte (black curve) are highlighted with vertical lines, representing two specific examples: 1M LiPF$_6$ in EMC and 1M LiFSI in FEC. Similarly, density variations for the binder (gray curve) indicate densities of Na-CMC and PVDF. The gradient of the electrolyte density variation is notably steeper than that of the binder variation, which can be attributed to its higher mass fraction in the battery cell's total weight. Even a slight reduction in

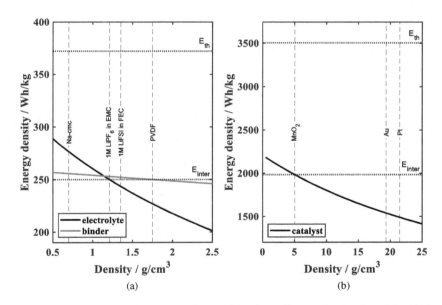

**Figure 3.5.** Energy density curves of (a) an LiB depending on the density of the binder and the liquid electrolyte used in the cell, (b) an Li-air battery depending on the density of the catalyst used in the cathode. Dotted horizontal lines indicate the theoretical energy density $E_{th}$ and the intermediate energy density $E_{inter}$ for the cell configuration specified in Table 3.2(1).

electrolyte density results in a substantial increase in energy density, underscoring the critical role of electrolyte choice in battery production.

The LiB example demonstrates how energy density changes with variations in (a) weight/volume fractions, (b) layer thicknesses, and (c) density of passive components. These principles are applicable to all other cell chemistries examined. Therefore, the subsequent discussion will focus exclusively on the distinctive features of each specific cell chemistry and their impact on energy density.

### (2) Lithium–metal battery

In this section, LiM batteries are categorized into two types based on the electrolyte used: (a) liquid electrolyte-based LiM batteries and (b) solid electrolyte-based LiM batteries, referred to as ASSBs.

#### a. Liquid electrolyte-based lithium–metal battery

In this particular use case, LiM differs from LiB solely in the design of passive cell components within the anode (Figure 3.6). As indicated by the reaction equation in Table 3.1(2), the cathode serves as the reservoir for lithium, while the lithium metal anode forms reversibly *in situ* during the initial charging process. However, for reasons of stability, practical applications often include an excess of lithium ($Li_{excess}$), as illustrated in Figure 3.6 and emphasized with a dashed outline.

Since the excess lithium does not participate in the electrochemical reaction, it is considered an additional passive cell component. Consequently, an additional BIRKE-summand, $f_{P9} Li_{excess}$, is included in the overall reaction equation, as depicted in Table 3.5(2a). The volumetric distribution of cell components contributes to the overall reaction equation depicted in Table 3.5(2a) and forms the basis for calculating energy

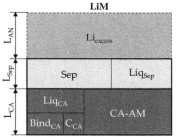

**Figure 3.6.** Volumetric distribution of cell components in an LiM battery.

density. Based on the cell composition detailed in Table 3.2(2a), the intermediate energy density $E_{inter}$ of an LiM is calculated to be 316 Wh/kg, thereby reducing the theoretical energy density $E_{th}$ by 40%. The cell composition of this LiM use case is identical to that of LiB, except that a 20 μm lithium foil is used as the anode.

To characterize the influence of the passive cell components on the energy density, Figure 3.7 shows the energy density curves depending on: (a) the volume fraction of varying passive cell components relative to CA-AM and (b) the layer thickness of the anode, cathode, and separator. The profiles of all the passive cathode components in Figure 3.6(a) are consistent with the LiB profiles. Additionally, the energy density trends observed with changes in separator and cathode thickness in Figure 3.6(b) align with those observed in the LiB.

The variation in anode thickness exhibits an inverse trend: The energy density decreases as the anode thickness increases. This observation pertains to the excess lithium layer, which does not participate in the electrochemical reaction and is thus considered a passive cell component. As the

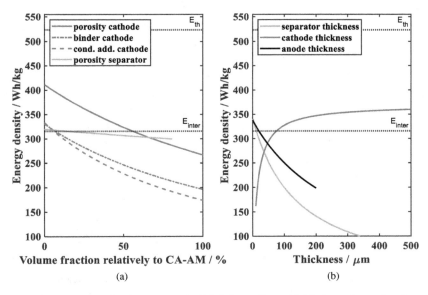

**Figure 3.7.** Energy density curves for LiM depending on: (a) the volume fraction of passive cell components relative to CA-AM, and (b) the layer thickness of separator, cathode, and anode. Dotted horizontal lines indicate the theoretical energy density $E_{th}$ and the intermediate energy density $E_{inter}$ for the cell configuration specified in Table 3.2(2a).

anode thickness increases, the proportion of passive components in the cell system also increases, thereby reducing the energy density.

*b. All-solid-state battery*

In the ASSB, the liquid electrolyte in both the cathode and the separator is completely replaced by an SSE (Figure 3.8). The SSE within the separator layer is referred to as the separator electrolyte ($SSE_{SSE}$), while the SSE contained in the composite cathode is termed the catholyte ($SSE_{CA}$). Additionally, a binder ($Bind_{SSE}$) is required in the separator layer to ensure adequate contacting and processability. The replacement of all the liquid components with SSE results in the inclusion of BIRKE-summands $f_{P4}$ $Bind_{SSE}$, $f_{P5}$ $SSE_{SSE}$, and $f_{P6}$ $SSE_{CA}$ in the overall reaction equation in Table 3.5(2b). All other passive cell components remain identical to those in LiM.

In the field of ASSBs, three main types are distinguished based on the electrolyte used: (1) sulfide-based ASSBs, (2) oxide-based ASSBs, and (3) polymer-based ASSBs. There has also been research exploring hybrid approaches to combine the advantages of these individual technologies. For the purposes of this chapter, sulfide-based ASSBs, abbreviated as "sulfide ASSB," and oxide-based ASSBs, abbreviated as "oxide ASSB," have been selected as representative examples for energy assessment.

i. Sulfide ASSB

Sulfide ASSBs, including crystalline structures such as $Li_6PS_5X$ (X = Cl, Br, I) of the argyrodite type, are distinguished by their exceptionally high ionic conductivity. For instance, $Li_6PS_5Cl$ demonstrates a conductivity of approximately $1.9 \times 10^{-3}$ S/cm, approaching the conductivity levels of liquid electrolytes, which are typically around $2 \times 10^{-2}$ S/cm [33].

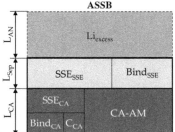

**Figure 3.8.** Volumetric distribution of cell components in an ASSB.

To assess the impact of the SSE on energy density, $Li_6PS_5Cl$ (LiPSCl) was chosen to function as both the SSE within the separator and as the catholyte. The cell composition and electrode formulations used are inspired by Ates *et al.* [34] and are listed in Table 3.2(2a i.).

The intermediate energy density $E_{inter}$ for this exemplary sulfide ASSB use case is calculated to be 293 Wh/kg. Consequently, the passive cell components reduce the theoretical energy density $E_{th}$ by 44%, clearly demonstrating the negative impact of the SSE on energy density. This reduction is attributed to the higher density of LiPSCl (1.64 g/cm³) compared to the liquid electrolyte (1.21 g/cm³ for $LiPF_6$ in EMC) used in LiM. To illustrate the impact of variations in passive cell components on energy density, Figure 3.6 presents energy density curves based on: (a) the volume fraction of different passive cell components relative to CA-AM (Figure 3.9(a)) and 3.9(b) the thickness of the cell layers, including the anode, cathode, and SSE separator (Figure 3.9(b)). Notably, in Figure 3.9(a), the energy density exhibits an almost horizontal trend despite the increasing volume fraction of

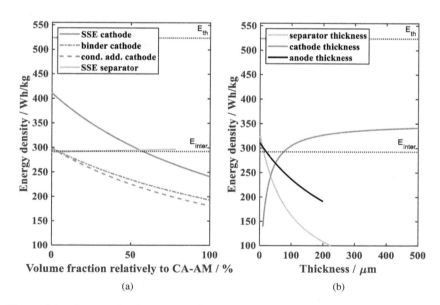

**Figure 3.9.** Energy density curves of a sulfide ASSB depending on: (a) the volume fraction of passive cell components relative to CA-AM, and (b) the layer thickness of separator, cathode, and anode. For ASSBs, the separator thickness equals the SSE layer thickness. Dotted horizontal lines indicate the theoretical energy density $E_{th}$ and the intermediate energy density $E_{inter}$ for the cell configuration specified in Table 3.2(2b i.).

the SSE. This behavior is attributed to the density of the SSE material itself. With a density of 1.64 g/cm³ for LiPSCl, the SSE is slightly lighter than the separator binder PVDF, which has a density of 1.75 g/cm³ and is the only other component in the separator. As the volume fraction of the SSE increases in the separator, there is a corresponding reduction in the volume fraction of the denser binder material. Consequently, the overall weight of the separator decreases, along with the weight of passive cell components relative to active components, leading to an enhanced energy density.

ii. Oxide ASSB

When compared to sulfide-based SSEs, oxide-based SSEs such as garnet-structured $Li_7La_3Zr_2O_{12}$ (LLZO) exhibit greater stability in the presence of ambient moisture and demonstrate inertness toward the lithium metal anode [35]. However, a significant drawback of oxide-based SSEs is their high density, stemming from the presence of heavy elements such as lanthanum (La, 138.9 g/mol) and zirconium (Zr, 91.22 g/mol). For instance, LLZO has a density of 5.01 g/cm³, more than three times that of LiPSCl, leading to a substantial increase in the overall weight of the cell. To precisely assess the impact of LLZO's high density on energy density, the same cell composition as that used for the sulfide ASSB was employed for comparison purposes, except that the SSE in both the separator and cathode was replaced with LLZO (Table 3.2(2b ii.)). The overall reaction equation also remains unchanged (Table 3.5(2b)). The intermediate energy density $E_{inter}$ amounts to 261 Wh/kg, which is notably lower compared to the sulfide ASSB. In this oxide ASSB use case, the passive cell components consequently decrease the theoretical energy density $E_{th}$ by 50%.

The substantial weight contribution of the LLZO SSE is also evident in Figure 3.10(a) and 3.10(b). Unlike the sulfide ASSB, the energy density of the oxide ASSB increases as the SSE content in the separator rises. This phenomenon is attributed to LLZO's higher density compared to the separator binder. It also explains the sharp decline in the energy density curve of the catholyte (SSE cathode) as the SSE content decreases. In summary, achieving maximum usable energy density with an oxide-based SSE necessitates minimizing the electrolyte content in both the separator and cathode, as well as reducing the separator thickness to the minimal possible level. The development of ASSBs is primarily driven by the goal of realizing a stable

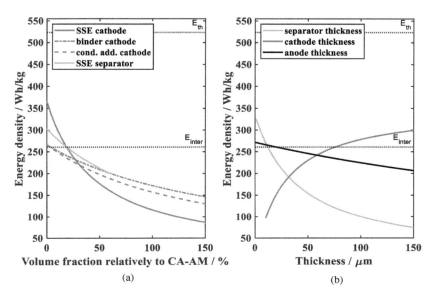

**Figure 3.10.** Energy density curves of an oxide ASSB depending on: (a) the volume fraction of passive cell components relative to CA-AM, and (b) the layer thickness of separator, cathode, and anode. For ASSBs, the separator thickness equals the SSE layer thickness. Dotted horizontal lines indicate the theoretical energy density $E_{th}$ and the intermediate energy density $E_{inter}$ for the cell configuration specified in Table 3.2(2b ii.).

lithium metal anode and the associated high energy density. However, the calculated results indicate that this motivation could be somewhat dampened by the significant weight contribution of the SSE. Therefore, an important question arises: Up to what SSE content is ASSB technology advantageous, with an energy density greater than that of LiB?

To address this question, Figure 3.11 presents the intermediate energy density $E_{inter}$ of the selected ASSBs, LiB, and LiM as a function of separator thickness. In ASSBs, the separator thickness corresponds to the thickness of the separator electrolyte layer, thereby indicating the solid electrolyte content in the battery. For the sulfide ASSB, the energy density falls below that of LiB at a maximum separator thickness of 70 $\mu$m. In contrast, for the oxide ASSB, this critical separator thickness is approximately 15 $\mu$m, which is significantly lower. Achieving such thin SSE layers poses substantial challenges. This insight prompts the question of whether, particularly for oxide ASSBs, the pursuit of higher "energy density" has reached its limits.

74   K. Schad & K. P. Birke

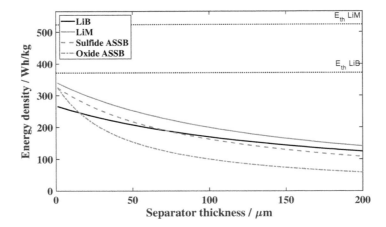

**Figure 3.11.** Influence of the separator thickness on the intermediate energy density $E_{inter}$ of different cell types: LiB, LiM, sulfide ASSB, and oxide ASSB. For ASSBs, the separator thickness equals the SSE layer thickness. Dotted horizontal lines indicate the theoretical energy density $E_{th}$ and the intermediate energy density $E_{inter}$ for the cell configuration for LiB (Table 3.2(1)) and LiM (Table 3.2(2a)).

### (3) Lithium–sulfur battery

Another lithium-based cell chemistry using metallic lithium as the anode is the Li–S battery. One of the primary challenges in Li–S batteries is achieving adequate electrical conductivity in the sulfur cathode. Given the substantial impact on cathode composition, discussed previously in Section 3.2.1. two representative design use cases are considered for assessing the energy density of the Li–S battery: (a) Li–S with excess conductive carbon in the cathode (Li–S C excess) and (b) Li–S with a conductive polymer matrix in the cathode (Li–S polymer matrix). These variants differ solely in the composition of the passive cathode components.

### a. Li–S C excess

Figure 3.12 schematically depicts the volumetric distribution of active and passive cell components in the Li–S C excess battery. Since conductive carbon serves as the conductive additive across all considered cell chemistries, the passive cell components in the Li–S C excess battery remain consistent with those of LiM (Figure 3.6). Consequently, the resulting overall reaction equation framework shown in Table 3.5(3a) is identical to

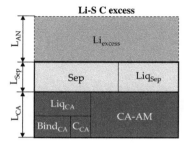

**Figure 3.12.** Volumetric distribution of cell components in an Li–S battery using excess conductive carbon ($C_{CA}$) to enhance cathodic conductivity.

that of LiM. The cell composition and electrode recipes are derived from Cheon et al. [9] and detailed in Table 3.2(3a). The intermediate energy density $E_{inter}$ is calculated to be 1189 Wh/kg. In this Li–S C excess use case, the passive cell components reduce the theoretical energy density $E_{th}$ by 54%.

To illustrate the influence of variations in passive cell components on energy density, Figure 3.13 shows energy density curves depending on: (a) the volume fraction of varying passive cell components relative to CA-AM and (b) the thickness of the cell layers including the anode, cathode, and separator. The observed energy density trends align with those of the LiM battery. A notable difference is a relatively shallow gradient of the energy density curve as a function of cathode thickness in Figure 3.13(b) compared to previous NMC-based cathodes. This behavior is attributed to the high carbon content $C_{CA}$ (27.3 wt.%) in the cathode of the Li–S C excess battery. With increasing cathode thickness, the proportionate increase in the weight of the passive cell component partially offsets the energy density gains achieved by the concurrent increase in CA-AM content.

*b. Li–S polymer matrix*
By incorporating a polymer matrix as an additional conductivity additive in the cathode, the Li–S polymer matrix battery introduces an additional passive component (Figure 3.14). This results in the inclusion of an additional BIRKE-summand $f_{P10}$ $Matx_{CA}$ in the overall reaction equation shown in Table 3.5(3b). Based on the work of Xiao et al. [10], the exemplary cell composition and electrode recipes detailed in Figure 3.2(3b) were used to calculate the intermediate energy density $E_{inter}$ as 1020 Wh/kg

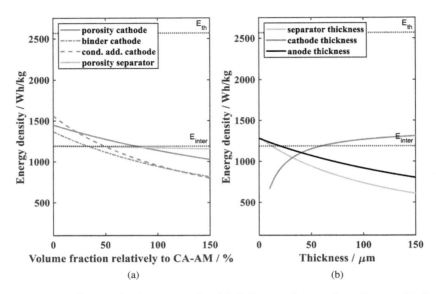

**Figure 3.13.** Energy density curves of an Li–S C excess battery depending on: (a) the volume fraction of passive cell components relative to CA-AM, and (b) the layer thickness of separator, cathode, and anode. Dotted horizontal lines indicate the theoretical energy density $E_{th}$ and the intermediate energy density $E_{inter}$ for the cell configuration specified in Table 3.2(3a).

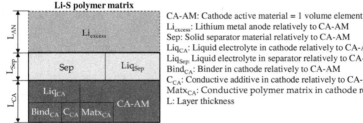

CA-AM: Cathode active material = 1 volume element
$Li_{excess}$: Lithium metal anode relatively to CA-AM
Sep: Solid separator material relatively to CA-AM
$Liq_{CA}$: Liquid electrolyte in cathode relatively to CA-AM
$Liq_{Sep}$: Liquid electrolyte in separator relatively to CA-AM
$Bind_{CA}$: Binder in cathode relatively to CA-AM
$C_{CA}$: Conductive additive in cathode relatively to CA-AM
$Matx_{CA}$: Conductive polymer matrix in cathode relatively to CA-AM
L: Layer thickness

**Figure 3.14.** Volumetric distribution of the cell components in an Li–S battery using a polymer matrix ($Matx_{CA}$) to enhance cathodic conductivity.

for the Li–S polymer matrix battery. As anticipated, employing a polymer matrix as an additional passive component results in a further decrease in energy density compared to the Li–S C excess cell system. In this Li–S polymer matrix battery use case, the passive cell components reduce the theoretical energy density $E_{th}$ by 60%.

The energy density effect of varying the proportion of the polymer matrix relative to the CA-AM volume is depicted in Figure 3.15(a). Similar to other passive cell components, the energy density of the Li–S polymer matrix battery decreases with an increase in the polymer matrix content. The variation in cathode thickness exhibits a comparable impact on energy density, as observed in the Li–S C excess battery. Despite containing only 49.6 wt.% CA-AM, the cathode exhibits a high proportion of passive cell components. Therefore, increasing cathode thickness results not only in more CA-AM but also in a significant rise in the proportion of passive components. This dual effect diminishes the positive energy density impact and contributes to the flattened curve observed in Figure 3.15(b).

The results from the Li–S battery depicted in Figures 3.13 and 3.15 demonstrate that, even when assuming maximum proportions of passive components and minimizing cathode thickness, the intermediate energy density $E_{inter}$ remains above 500 Wh/kg, thus exceeding the energy density of LiM (338 Wh/kg).

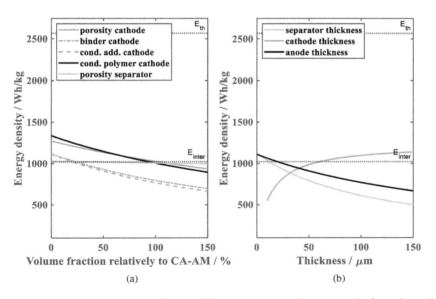

**Figure 3.15.** Energy density of an Li–S battery using a polymer matrix for enhanced conductivity depending on: (a) the volume fraction of passive cell components relative to CA-AM, and (b) the layer thickness of separator, cathode, and anode. Dotted horizontal lines indicate the theoretical energy density $E_{th}$ and the intermediate energy density $E_{inter}$ for the cell configuration specified in Table 3.2(3b).

However, it is important to acknowledge that these use cases are based on idealized assumptions. Currently, the Li–S technology faces significant challenges that hinder the achievable energy density at the cell level. The primary factors directly influencing energy density include poor sulfur utilization efficiency due to suboptimal diffusion kinetics and low electrical conductivity of sulfur, necessitating excess sulfur. Additionally, the dissolution of polysulfides poses a significant issue, leading to capacity loss over multiple cycles.

The energy density $E_{inter}$ considered here should therefore be interpreted as an idealized value that demonstrates what would theoretically be possible if the above-mentioned challenges were neglected.

### (4) Lithium–air battery

Due to the necessity of a catalyst in the cathode, the Li–air battery includes $Cat_{CA}$ as an additional passive component, alongside the standard passive components $Bind_{CA}$, $Ca_{CA}$, and $Liq_{CA}$ (Figure 3.16). Consequently, the overall reaction equation is expanded to include the BIRKE-summand $f_{P10} \, Cat_{CA}$, as outlined in Table 3.5(4).

The Li–air battery occupies a unique position compared to all other considered cell chemistries due to the phase transition of the solid cathode discharge product, $Li_2O_2$, to gaseous $O_2$ during the charging process. The gas diffusion cathode facilitates a continuous flow of $O_2$ during battery operation. Unlike other chemistries, where the volume fraction of active material is determined by weighing a defined mass fraction into the electrode slurry, in Li–air batteries, it is determined by the cathode's porosity. The pores within the cathode are filled with both liquid electrolyte and $O_2$, with their distribution changing during charging and discharging. The phase transformation and dynamic conditions in the pores require the

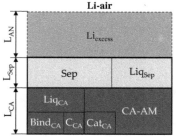

**Figure 3.16.** Volumetric distribution of cell components in an Li–air battery.

following simplifying assumptions for the applicability of the $E_{inter}$ calculation approach in the air cathode: The mass fraction of the discharge product $Li_2O_2$ is assumed to be constant at 70 wt.% because it is only formed during cell operation and not part of the weighted-in components [2]. The recipe for the weighed-in components is based on the work of Mizuno et al. [10], with $MnO_2$ selected as the catalyst material. Additionally, the porosity in the discharged state is approximated to be 20 vol.% [2]. In the discharged state, the residual air within the pores is considered negligible. Consequently, the volume fraction of the liquid electrolyte is assumed to be equivalent to the porosity.

Based on these assumptions and the selected Li–air use case (Table 3.2(4)), the intermediate energy density $E_{inter}$ is calculated to be 1984 Wh/kg. In this Li–air battery use case, the passive cell components therefore reduce the theoretical energy density $E_{th}$ by 43%. Figure 3.17(a) illustrates the energy density curves with varying fractions of the different passive cathode components.

The curves of the passive cathode components – binder, electrolyte, conductive additive, and catalyst – exhibit limited variation within the

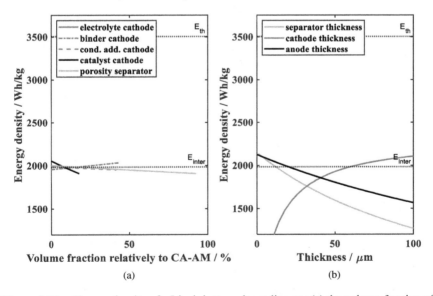

**Figure 3.17.** Energy density of a Li–air battery depending on: (a) the volume fraction of passive cell components relative to CA-AM, and (b) the layer thickness of separator, cathode, and anode. Dotted horizontal lines indicate the theoretical energy density $E_{th}$ and the intermediate energy density $E_{inter}$ for the cell configuration specified in Table 3.2(4).

volume range relative to the cathode. This constraint arises because the mass fraction of CA-AM is assumed to be constant, thereby limiting the volume of the passive cell components. This constant CA-AM fraction contributes to the low gradients of the energy density curves, as changes in the weight fraction of a passive component only affect the fractions of the other passive components, not the CA-AM fraction. Notably, variations in the catalyst fraction significantly influence energy density due to the catalyst's comparatively high density ($MnO_2$, 5.03 g/cm$^3$). To underscore the significant role played by the catalyst, the energy density curve over the density of the catalyst is illustrated in Figure 3.5. Vertical dotted lines highlight three widely used catalyst materials: $MnO_2$, platinum (Pt), and gold (Au). For example, using heavy Pt (21.45 g/cm$^3$) as a catalyst in the air cathode reduces the energy density by 500 Wh/kg compared to using an $MnO_2$ catalyst.

In conclusion, despite its passive cell components, the Li–air battery offers the potential for remarkably high energy densities. However, the practically usable energy density is currently significantly lower than the theoretically calculated 1984 Wh/kg. This discrepancy is primarily due to the extremely challenging complete conversion of $O_2$ to $Li_2O_2$, which leads to undesirable side reactions causing irreversible lithium consumption and the formation of additional passive cell components. Furthermore, the three-phase interface problem, particularly during load changes, contributes to stability issues and a reduction in capacity.

(5) **Sodium-ion battery**
With two intercalation-based electrodes, the structure of the SiB in terms of its passive cell components is identical to that of the LiB (Figure 3.18). Consequently, the overall reaction equation for an SiB also corresponds to

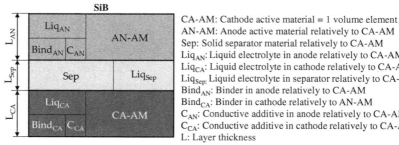

**Figure 3.18.** Volumetric distribution of the cell components in an SiB.

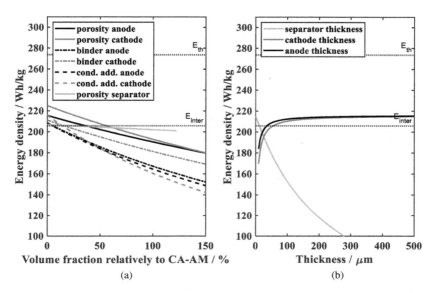

**Figure 3.19.** Energy density of an SiB depending on: (a) the volume fraction of passive cell components relative to CA-AM, and (b) the layer thickness of separator, cathode, and anode. Dotted horizontal lines indicate the theoretical energy density $E_{th}$ and the intermediate energy density $E_{inter}$ for the cell configuration specified in Table 3.2(5).

that of an LiB (Table 3.5(5)). To calculate the intermediate energy density $E_{inter}$ of an SiB, the cell composition proposed by Laufen et al. [3] was utilized, as shown in Table 3.5(5). It should be noted that HC, commonly used as an AN-AM material in SiBs, was approximated based on the reaction characteristics and material properties similar to graphite. In this use case, $E_{inter}$ is calculated to be 206 Wh/kg, indicating a 25% reduction from the theoretical energy density $E_{th}$ due to the passive cell components.

The energy density profiles, as shown in Figure 3.19(a) and 3.19(b), respectively, vary depending on the volume fractions of passive cell components and layer thicknesses. These profiles exhibit similar behavior to those observed in LiBs, rendering a detailed discussion of the results unnecessary.

### (6) Zinc battery

As previously introduced, two representative examples, the ZiB and the ZaB, are considered for evaluating the energy density of ZBs.

In both cases, the cell architecture concerning passive components is analogous to that of LiM batteries, except that, in the case of ZBs, the

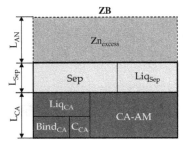

**Figure 3.20.** Volumetric distribution of cell components in a ZB.

lithium anode is replaced with zinc (Figure 3.20). Similar to the excess lithium in LiM batteries, ZBs are operated with excess zinc, termed $Zn_{excess}$, resulting in the BIRKE-summand $f_{p9}\, Zn_{excess}$.

*a. Zinc-ion battery*

In the intercalation-based ZiB, no electrolyte participation is assumed, resulting in the overall reaction equation depicted in Table 3.5(6a). This equation demonstrates similarities to that representing LiM batteries, which is reflected in the energy density behavior illustrated in Figure 3.21(a) and 3.21(b).

To calculate the intermediate energy density $E_{inter}$, the exemplary cell composition was inspired by the works of Song *et al.* [4] and Innocenti *et al.* [31] (Table 3.2(6a)). Assuming an N/P ratio of 1.1 results in an anode thickness of approximately 40 μm. $E_{inter}$ is calculated to be 166 Wh/kg. In this ZiB use case, the passive cell components therefore reduce the theoretical energy density $E_{th}$ by 43%.

Notably, Figure 3.21(b) demonstrates a significant decline in energy density as the anode thickness increases. Zinc, with a density of 7.14 g/cm³, is over 13 times greater than that of lithium (0.534 g/cm³). Consequently, increasing the thickness of the zinc anode adds substantial weight, contributing to the observed decrease in energy density.

*b. Zinc aqueous electrolyte battery*

The ZaB involves electrolyte participation in the reaction, while the passive cell components, including their corresponding BIRKE-summands, remain identical to those in ZiB. The resulting overall reaction equation is depicted in Table 3.5(6b). Inspired by the works of Song *et al.* [4] and Innocenti *et al.* [31], a cell composition similar to that of ZiB was chosen

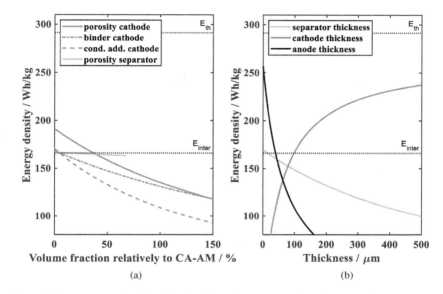

**Figure 3.21.** Energy density of a ZiB depending on: (a) the volume fraction of passive cell components relative to CA-AM, and (b) the layer thickness of separator, cathode, and anode. Dotted horizontal lines indicate the theoretical energy density $E_{th}$ and the intermediate energy density $E_{inter}$ for the cell configuration specified in Table 3.2(6a).

for comparability. ZiB and ZaB primarily differ in their electrolyte choice and the thickness of the zinc anode. For the ZaB, in which the electrolyte participates as a reactant, the pH of the electrolyte solution plays a crucial role. Consequently, a slightly acidic aqueous 1 M $ZnSO_4$ / 1 M $MnSO_4$ electrolyte solution is selected. The difference in anode thickness stems from ZaB's (616 mAh/g) cathode capacity being twice that of ZiB (308 mAh/g) [4]. An N/P ratio of 1.1 thus implies an anode thickness of approximately 80 μm for ZaB. The intermediate energy density $E_{inter}$ is calculated to be 157 Wh/kg. In this ZaB use case, the passive cell components reduce the theoretical energy density $E_{th}$ by 32%.

Since the energy density behavior is identical to that of ZiB concerning the variation of the proportion of passive cell components and the layer thicknesses (Figure 3.22), further discussion on this topic will not be provided here.

It is important to acknowledge that the actual energy density $E_{real}$ in practical applications of ZBs is lower than the calculated intermediate energy density $E_{inter}$. The use of an aqueous electrolyte precludes the use

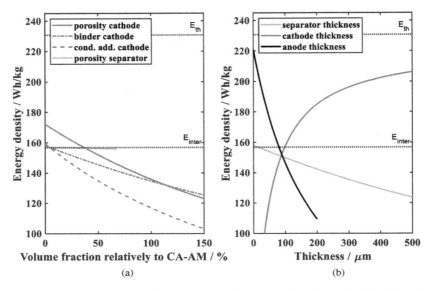

**Figure 3.22.** Energy density of a ZaB depending on: (a) the volume fraction of passive cell components relative to CA-AM, and (b) the layer thickness of separator, cathode, and anode. Dotted horizontal lines indicate the theoretical energy density $E_{th}$ and the intermediate energy density $E_{inter}$ for the cell configuration specified in Table 3.2(6b).

of aluminum (2.7 g/cm³) as a current collector, often requiring the substitution with significantly heavier materials such as stainless steel (Stainless Steel 1.4301, 7.9 g/cm³). As a result, the current collector concept of the ZB imposes a significantly higher weight contribution when compared to the aluminum current collector foils typically used as the cathode in lithium- and sodium-based cells. Furthermore, ZBs experience substantial zinc loss during operation, necessitating a high N/P ratio (~2) to ensure cycle stability, resulting in thicker zinc anodes and additional weight.

(7) **Lead–acid battery**

The classical Pb battery, as depicted in Figure 3.23, shares analogous passive components in the anode and cathode with those of the previously discussed LiB and SiB. However, the main difference lies in the mechanism of charge storage. While LiB and SiB are based on an intercalation mechanism, the Pb battery operates through a conversion process, with the electrolyte acting as a reactant at both the anode and the cathode. Due

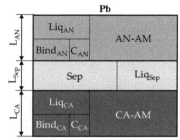

Figure 3.23. Volumetric distribution of cell components in a Pb battery.

to the electrolyte's direct involvement in the reaction, the fundamental principle of the Pb battery resembles that of the ZaB, leading to the overall reaction equation depicted in Table 3.5(7).

With the exemplary cell composition according to Sulzer et al. [36] (Table 3.2(7)), the intermediate energy density $E_{inter}$ is calculated to be 123 Wh/kg. In this Pb battery use case, the passive cell components therefore reduce the theoretical energy density $E_{th}$ by 26%.

The energy density profiles in Figure 3.24(a) and 3.24(b), depending on varying volume fractions of the passive cell components and layer thicknesses, respectively, exhibit behavior similar to that observed in LiB and SiB, thereby making a detailed discussion of the results unnecessary. However, two anomalies should be mentioned: First, due to the high density of Pb ($PbO_2$, 9.34 g/cm³ and Pb, 11.34 g/cm³), it occupies significantly less volume compared to the lighter passive cell components. As a result, in Figure 3.24(a), the volume fractions of the passive cell components are relatively shifted toward higher proportions compared to CA-AM. Therefore, for example, the volume fraction of the liquid electrolyte in the cathode is 1.7 times higher than that of CA-AM. Second, the Pb battery is characterized by high layer thicknesses of the electrodes and separator. For example, the cathode thickness in the Pb use case is 1250 $\mu$m, which is more than 15 times thicker than the cathode layer in an LiB. This necessity arises from the high porosity of the electrodes (e.g., cathode porosity: 57 vol.%) due to the significant electrolyte requirement. Reducing the cathode thickness, for example, to 250 $\mu$m would lead to a drastic decrease in the ratio between active material and passive components in the cathode, thereby reducing the energy density to less than 100 Wh/kg.

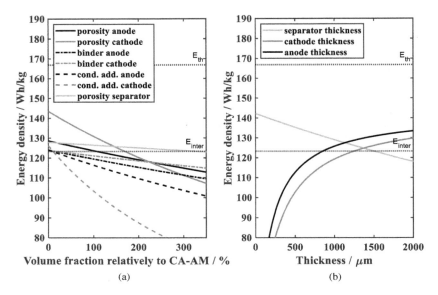

**Figure 3.24.** Energy density of a Pb battery depending on: (a) the volume fraction of passive cell components relative to CA-AM, and (b) the layer thickness of separator, cathode, and anode. Dotted horizontal lines indicate the theoretical energy density $E_{th}$ and the intermediate energy density $E_{inter}$ for the cell configuration specified in Table 3.2(7).

As with the ZB, the current collector concept of the Pb battery significantly differs from those used in lithium- and sodium-based cells. The Pb battery employs grid-like structures made of lead or lead alloys to provide mechanical stabilization for the active materials. Due to the high density of Pb (11.34 g/cm³), the current collector concept has a significantly higher weight contribution compared to the aluminum (Al, 2.7 g/cm³) and copper (Cu, 8.96 g/cm³) current collector foils used in lithium- and sodium-based cells. Consequently, the actual energy density $E_{real}$ of a Pb battery is considerably lower than the calculated intermediate energy density $E_{inter}$. Additionally, the requirement for excess electrolyte and limited SOC usability further reduces the real energy density.

### 3.2.3 Comparison of the theoretical and intermediate energy densities

Finally, insights from the energy density considerations of different cell chemistries are compared. Figure 3.25 depicts the theoretical energy

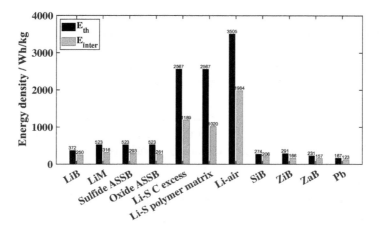

**Figure 3.25.** Comparison of the theoretical energy density $E_{th}$ and the intermediate energy density $E_{inter}$ of the selected cell chemistries.

density $E_{th}$ and the calculated intermediate energy density $E_{inter}$ for all the considered cell chemistries.

The influence of passive cell components on energy density varies significantly among different cell chemistries. The following sequence of the considered cell chemistry use cases emerges regarding the loss of energy density due to the inclusion of passive components: SiB (25%) < Pb (26%) < ZaB (32%) < LiB (33%) < LiM (40%) < ZiB/Li–air (43%) < Sulfide ASSB (44%) < Oxide ASSB (50%) < Li–S C excess (54%) < Li–S polymer matrix (60%).

The intermediate energy density $E_{inter}$ reveals significant trends regarding the limitations of various cell systems. In ASSBs, the heavy solid electrolyte primarily contributes to the energy density loss. For the Li–S battery, the required excess of conductive additives plays a key role in reducing energy density. In the Li–air battery, the heavy catalyst is the dominant factor decreasing energy density, while in zinc-based batteries, the zinc anode has this impact. Similarly, in Pb batteries, the high electrolyte content is the main contributor to energy density loss.

Since exemplary use cases and idealized assumptions were used to define the cell composition and since the current collectors and casing components were not included in the calculations, $E_{inter}$ represents only an intermediate step toward the real usable energy density $E_{real}$ (Figure 3.1).

Considering zinc-based systems ZiB/ZaB and the Pb battery, the "heavy" current collector concepts are not transferable to other cell chemistries. This results in a notable disparity between $E_{inter}$ and $E_{real}$. While $E_{inter}$ serves as an intermediate measure, a meaningful comparison of the real energy densities is only feasible within a casing and current collector reference system.

The casing and current collector concept of LiB can be extended to LiM and ASSBs, establishing a common reference system in this scenario. Consequently, the intermediate comparison becomes highly informative, facilitating a technology evaluation derived from the calculated $E_{inter}$. The results of ASSBs indicate that the original technology driver, energy density, remains viable only under ideal conditions, such as the application of very thin and therefore highly challenging solid electrolyte layers. The profitability of solid-state batteries, particularly regarding their energy density, is thus constrained within these limits.

The landscape of battery storage is diverse and lacks standardized norms, making the choice of the reference system essential for comparing different technologies. This chapter introduced a flexible calculation approach for the intermediate energy density $E_{inter}$, which considers passive cell components across various cell chemistries. To facilitate the applicability of this approach, cell casing and current collectors were deliberately excluded from the calculations. As a result, the calculated energy density $E_{inter}$ serves as an intermediate measure, enabling the assessment of selected passive cell components based on their impact on energy density.

# References

[1] D. Karabelli and K. P. Birke, Feasible energy density pushes of Li-metal vs. Li-ion cells, *Applied Sciences*, 11(16), 7592, 2021, doi:10.3390/app11167592.

[2] P. G. Bruce, S. A. Freunberger, L. J. Hardwick *et al.*, Li-O$_2$ and Li-S batteries with high energy storage, *Nature Materials*, 11(1), 19–29, 2011, doi:10.1038/nmat3191.

[3] H. Laufen, S. Klick, H. Ditler *et al.*, Multi-method characterization of a commercial 1.2 Ah sodium-ion battery cell indicates drop-in potential, *Cell Reports Physical Science*, 5(1), 101945, 2024, doi:10.1016/j.xcrp.2024.101945.

[4] J. Song, K. Xu, N. Liu *et al.*, Crossroads in the renaissance of rechargeable aqueous zinc batteries, *Materials Today*, 45, 191–212, 2021, doi:10.1016/j.mattod.2020.12.003.

[5] N. Wang, X. Qiu, J. Xu *et al.*, Cathode materials challenge varied with different electrolytes in zinc batteries, *ACS Materials Letters*, 4(1), 190–204, 2022, doi:10.1021/acsmaterialslett.1c00499.

[6] D. Linden, T. B. Reddy (Eds.), *Handbook of Batteries*, 3rd ed., McGraw-Hill, New York, NY, 2002, ISBN 0071359788.

[7] A. Manthiram, S.-H. Chung, C. Zu, Lithium-sulfur batteries: Progress and prospects, *Advanced Materials*, 27(12), 1980–2006, 2015, doi:10.1002/adma.201405115.

[8] Y. Huang, L. Lin, C. Zhang *et al.*, Recent advances and strategies toward polysulfides shuttle inhibition for high-performance Li-S batteries, *Advanced Science*, 9(4), e2106004, 2022, doi:10.1002/advs.202106004.

[9] S.-E. Cheon, K.-S. Ko, J.-H. Cho *et al.*, Rechargeable lithium sulfur battery, *Journal of the Electrochemical Society*, 150(6), A800, 2003, doi:10.1149/1.1571533.

[10] L. Xiao, Y. Cao, J. Xiao *et al.*, A soft approach to encapsulate sulfur: Polyaniline nanotubes for lithium-sulfur batteries with long cycle life, *Advanced Materials*, 24(9), 1176–1181, 2012, doi:10.1002/adma.201103392.

[11] P. Tan, H. R. Jiang, X. B. Zhu *et al.*, Advances and challenges in lithium-air batteries, *Applied Energy*, 204, 780–806, 2017, doi:10.1016/j.apenergy.2017.07.054.

[12] F. Mizuno, S. Nakanishi, Y. Kotani *et al.*, Rechargeable Li-air batteries with carbonate-based liquid electrolytes, *Electrochemistry*, 78(5), 403–405, 2010, doi:10.5796/electrochemistry.78.403.

[13] C. Xu, B. Li, H. Du *et al.*, Energetic zinc ion chemistry: The rechargeable zinc ion battery, *Angewandte Chemie International Edition*, 51(4), 933–935, 2012, doi:10.1002/anie.201106307.

[14] B. Lee, H. R. Seo, H. R. Lee *et al.*, Critical role of pH evolution of electrolyte in the reaction mechanism for rechargeable zinc batteries, *ChemSusChem*, 9(20), 2948–2956, 2016, doi:10.1002/cssc.201600702.

[15] L. Li, T. K. A. Hoang, J. Zhi *et al.*, Functioning mechanism of the secondary aqueous Zn-$\beta$-MnO2 battery, *ACS Applied Materials & Interfaces*, 12(11), 12834–12846, 2020, doi:10.1021/acsami.9b22758.

[16] D. Chao, W. Zhou, C. Ye *et al.*, An electrolytic Zn–$MnO_2$ battery for high-voltage and scalable energy storage, *Angewandte Chemie*, 131(23), 7905–7910, 2019, doi:10.1002/ange.201904174.

[17] M. Iturrondobeitia, O. Akizu-Gardoki, O. Amondarain *et al.*, Environmental impacts of aqueous zinc ion batteries based on life cycle assessment, *Advanced Sustainable Systems*, 6(4), 2022, doi:10.1002/adsu.202100308.

[18] A. Bayaguud, Y. Fu, C. Zhu, Interfacial parasitic reactions of zinc anodes in zinc ion batteries: Underestimated corrosion and hydrogen evolution reactions and their suppression strategies, *Journal of Energy Chemistry*, 64, 246–262, 2022, doi:10.1016/j.jechem.2021.04.016.

[19] C.-H. Chen, F. Brosa Planella, K. O'Regan *et al.*, Development of experimental techniques for parameterization of multi-scale lithium-ion battery models, *Journal of the Electrochemical Society*, 167(8), 80534, 2020, doi:10.1149/1945-7111/ab9050.

[20] S. Zhong, M. Lai, W. Yao *et al.*, Synthesis and electrochemical properties of LiNi0.8CoxMn0.2-xO2 positive-electrode material for lithium-ion batteries, *Electrochimica Acta*, 212, 343–351, 2016, doi:10.1016/j.electacta.2016.07.040.

[21] The Materials Project, mp-1153: Li2S (Cubic, Fm-3m, 225), Available online: https://next-gen.materialsproject.org/materials/mp-1153?formula=Li2S (accessed on 5 July 2024).

[22] The Materials Project, $Li_2O_2$ mp-1180619, Available online: https://next-gen.materialsproject.org/materials/mp-1180619?formula=Li2O2 (accessed on 5 July 2024).

[23] The Materials Project, $MnO_2$ mp-644514, Available online: https://next-gen.materialsproject.org/materials/mp-644514?formula=MnO2 (accessed on 5 July 2024).

[24] The Materials Project, $MnSO_4$ mp-22554, Available online: https://next-gen.materialsproject.org/materials/mp-22554?formula=MnSO4#literature_references (accessed on 5 July 2024).

[25] The Materials Project, $PbSO_4$ mp-3472, Available online: https://next-gen.materialsproject.org/materials/mp-3472?formula=PbSO4 (accessed on 5 July 2024).

[26] The Materials Project, $PbO_2$ mp-20725, Available online: https://next-gen.materialsproject.org/materials/mp-20725?formula=PbO2#summary (accessed on 5 July 2024).

[27] H. O. Pierson, *Handbook of carbon, graphite, diamond, and fullerenes: Properties, processing, and applications*, Noyes Publications, Park Ridge, N.J, 1993, ISBN 0815517394.

[28] The Materials Project, $ZnSO_4$ mp-5126, Available online: https://next-gen.materialsproject.org/materials/mp-5126?formula=ZnSO4 (accessed on 5 July 2024).

[29] Materials Project, *Materials Data on Li6PS5Cl by Materials Project*, 2020.

[30] Materials Project, *Materials Data on Li7La3Zr2O12 by Materials Project*, 2020.

[31] A. Innocenti, D. Bresser, J. Garche *et al.*, A critical discussion of the current availability of lithium and zinc for use in batteries, *Nature Communications*, 15, 4068, 2024, doi:10.1038/s41467-024-48368-0.

[32]　Schwefelsäure 99.999% | Sigma-Aldrich, Available online: https://www. sigmaaldrich.com/DE/de/product/aldrich/339741 (accessed on 5 July 2024).

[33]　D. Karabelli, K. P. Birke, M. Weeber, A performance and cost overview of selected solid-state electrolytes: Race between polymer electrolytes and inorganic sulfide electrolytes, *Batteries*, 7(1), 18, 2021, doi:10.3390/batteries7010018.

[34]　T. Ates, M. Keller, J. Kulisch *et al.*, Development of an all-solid-state lithium battery by slurry-coating procedures using a sulfidic electrolyte, *Energy Storage Materials*, 17, 204–210, 2019, doi:10.1016/j.ensm.2018.11.011.

[35]　X. Yan, Z. Li, Z. Wen *et al.*, Li/Li7La3Zr2O12/LiFePO4 all-solid-state battery with ultrathin nanoscale solid electrolyte, *Journal of Physical Chemistry C*, 121(2), 1431–1435, 2017, doi:10.1021/acs.jpcc.6b10268.

[36]　V. Sulzer, S. J. Chapman, C. P. Please *et al.*, Faster lead-acid battery simulations from porous-electrode theory: Part I. Physical model, *Journal of the Electrochemical Society*, 166(10), A2363-A2371, 2019, doi:10.1149/2.0301910jes.

© 2025 World Scientific Publishing Company
https://doi.org/10.9789811282058_0004

# Chapter 4

# Energy Density versus Power Density, Lifespan, Safety, and Costs

**Sabri Baazouzi[*] and Kai Peter Birke[†]**

*Fraunhofer IPA, Nobelstrasse 12, Stuttgart, Germany*

*[*]sabri.baazouzi@outlook.de*

*[†]kai.peter.birke@ipa.fraunhofer.de*

This chapter delves into the evolving landscape of lithium-ion (Li-ion) battery cell formats utilized in the automotive industry: pouch, prismatic, and cylindrical cells. Each cell type is analyzed based on its casing technology, electrode design, and the resulting impact on energy and power density, safety, cost, and lifespan. A significant focus is placed on cylindrical cells, detailing their historical development and current trends. Cylindrical cells, particularly advancements from 18650 to 4680 cells, are explored in depth, highlighting their production efficiencies, enhanced energy densities, and the role of innovative design features such as tabless designs to mitigate inhomogeneities. The chapter provides a comparative analysis of cell capacities and energy densities, underscoring the progress achieved from 2010 to 2021. By examining 19 different cylindrical cells from leading manufacturers, the chapter offers insights into the generic characteristics and technological improvements driving the adoption of cylindrical cells in modern electric vehicles. This comprehensive study aims to illuminate the complexities and advancements in battery cell technology, contributing to a deeper understanding of their applications in the automotive sector.

## 4.1 Introduction: Overview of Cell Formats

In the automotive industry, three distinct types of lithium-ion (Li-ion) cells are utilized: pouch cells, prismatic cells, and cylindrical cells [1] (refer to Figure 4.1). These cells vary in both casing technology and the design and production of the electrode stack or jelly roll. Pouch cell casings are made of coated aluminum foil, while prismatic cells have a rigid metal casing. Similarly, cylindrical cells are encased in a rigid steel or aluminum housing. The selection of the cell type significantly influences the battery architecture, energy density, thermal management, safety, costs, and lifespan.

Pouch cells contain stacked electrodes, whereas prismatic cells use a flat jelly roll combination composed of one or more jelly rolls. Despite the production advantages offered by winding technology, there is a current trend toward adopting stacked electrodes in prismatic cells. Stacking eliminates the winding curves, leading to improved volume utilization and consequently higher energy density [2]. Moreover, the stacking process is associated with reduced mechanical stress on the electrode material compared to the flat winding technique [2]. Cylindrical cells, in contrast, exclusively employ jelly rolls. The advanced winding technique is a key benefit of cylindrical cells. With innovative cell designs, cylindrical cells are gaining popularity in the automotive field due to the reduction of inhomogeneities within the cell, which facilitates the development of large-format battery cells, such as the 4680 cells [3,4].

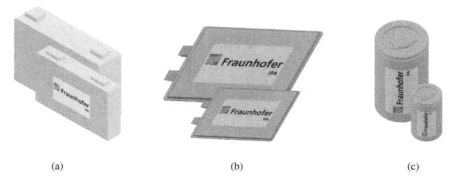

(a)  (b)  (c)

**Figure 4.1.** Formats of Li-ion battery cells: (a) prismatic cells, (b) pouch cells, (c) cylindrical cells.

## 4.2 Cell Properties: Cell Capacity and Energy Density

In this section, the properties of cell capacity and energy density will be explained and compared for all three cell formats based on the evaluation made by Link *et al.* [5].

### 4.2.1 *Cell capacity*

The capacity of a Li-ion battery cell quantifies the electric charge it can hold, typically measured in ampere-hours (Ah). This capacity is determined by integrating the current that the battery emits during discharge until it reaches a predetermined end-of-discharge voltage. Factors such as the cell's temperature, discharge rate (C-rate), and its calendrical age or number of completed charge/discharge cycles also influence a Li-ion battery cell's capacity. Manufacturers provide a nominal capacity based on specific conditions, often at a slow discharge rate and room temperature. However, the actual usable capacity may be lower than the nominal capacity due to inefficiencies, and it can diminish as the battery ages.

Between 2010 and 2021, the average capacity of cylindrical and pouch cells doubled, reaching approximately 70 Ah for pouch cells and 4.5 Ah for cylindrical ones. This significant increase in capacity, particularly notable after 2015, is primarily attributed to the introduction of new active materials that provide higher energy densities rather than to a substantial increase in cell size. The study by Link *et al.* [5] did not include Tesla's 4680 cell, which has a capacity of 22 Ah [6], representing a tenfold increase compared to the early 2010s. Prismatic cells, on the other hand, have a notably higher capacity than both cylindrical and pouch cells. Consequently, they have not seen a substantial capacity increase in recent years, largely due to concerns related to safety and temperature management.

### 4.2.2 *Energy density*

The energy density of Li-ion battery cells can be assessed from both gravimetric and volumetric perspectives.

The gravimetric energy density ($E_{grav}$) refers to the amount of energy stored per unit of weight, as defined in equation (4.1), where $m_{cell}$

represents the cell weight, $U$ denotes electrical voltage, and $I$ signifies electrical current. This parameter is especially critical for weight-sensitive applications such as electric vehicles (EVs) or portable electronic devices:

$$E_{\text{grav}} = \frac{1}{m_{\text{cell}}} \int UI \, dt. \tag{4.1}$$

The volumetric energy density $(E_{\text{vol}})$ refers to the amount of energy stored per unit volume, as defined in equation (4.2), where $V_{\text{cell}}$ represents the cell volume and $t$ denotes time. This parameter is crucial for applications where space is limited and compact battery design is essential:

$$E_{\text{grav}} = \frac{1}{m_{\text{cell}}} \int UI \, dt. \tag{4.2}$$

From 2010 to 2021, there was a notable upward trend in the energy density across all cell formats. Cylindrical cells showed a more gradual increase compared to pouch and prismatic cells, reaching 250 Wh/kg and 700 Wh/l by 2021, as illustrated in Figure 4.2(e) and 4.2(h). Pouch cells, shown in Figure 4.2(d) and 4.2(g), experienced a marked rise in energy density, especially from 2015 to 2018, surpassing 260 Wh/kg in 2021 and outperforming cylindrical cells. The volumetric energy density of pouch cells also rose to around 560 Wh/l. However, estimates by Link *et al.* may be conservative, as manufacturer data suggest potential maximums close to 670 Wh/l, which rivals that of cylindrical cells. Prismatic cells, as shown in Figure 4.2(c) and 4.2(f), exhibited slightly lower energy densities at the cell level but still demonstrated significant improvements. From 2019 onward, prismatic cells maintained average energy densities of 400–450 Wh/l, with NMC/NCA cells reaching approximately 550 Wh/l and LFP cells around 380 Wh/l by 2021. Some cells even surpassed the 650 Wh/l threshold, placing them on par with other cell formats in terms of peak energy density.

## 4.3 Deep Dive: Cylindrical Battery Cells

Cylindrical battery cells play a crucial role in the automotive sector and various other applications due to numerous advantages, such as simple production and high mechanical stability. However, conventional

Energy Density versus Power Density, Lifespan, Safety, and Costs 97

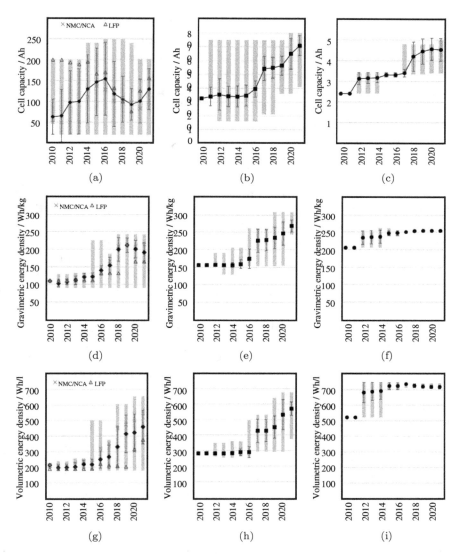

**Figure 4.2.** Evolution of cell capacity and gravimetric and volumetric energy densities of the three cells formats, prismatic, pouch, and cylindrical, between 2010 and 2021: (a)–(c) capacity trends, (d)–(f) gravimetric energy density improvements, and (g)–(i) volumetric energy density enhancements, each set corresponding to the three cell types in sequence.

*Source*: The subfigures were collected from Link *et al.* [5].

cylindrical cells have certain design drawbacks that limit their scalability in dimensions. On the other hand, scaling up cylindrical cells presents a promising approach to significantly reduce the number of passive components in a battery system. This scalability can contribute to increasing the energy density of EV batteries [7] and reducing production costs at the cell, module, and system levels [8] by simplifying production processes and minimizing the number of steps involved [9].

However, the limited mass and charge transport mechanisms in cylindrical cells with classical tab design pose numerous challenges due to inhomogeneities in various physical properties, such as temperature, current density, mechanical stress, state of charge (SoC), and particle concentration [9–12]. Waldmann *et al.* [13] demonstrated that the use of tabs in conventional cells significantly impacts their cyclic lifespan. The disadvantages of the classic tab design include safety concerns related to the risk of thermal runaway, limitations in lifespan due to accelerated and uneven aging, and efficiency losses caused by uneven utilization of electrode materials. Moreover, the design necessitates additional components, leading to production disadvantages such as increased production steps, discontinuities in the production chain, and increased quality issues.

Scalable cell designs are characterized by axially arranged electrical and thermal paths, independent of the electrode length. Such designs, known as "tabless" cells in the literature [6], significantly eliminate inhomogeneities in cylindrical battery cells, enabling the realization of large-format cells for future battery architectures. However, this advancement necessitates new processes in winding production and cell contacting to connect the jelly roll to the cell housing. In this chapter, we address the following research question: What cell designs and production methods are necessary to increase the volume of cylindrical battery cells, thereby reducing the number of components in a battery system and enhancing energy density?

### 4.3.1 *Cylindrical battery cells in the automotive sector*

The use of cylindrical cells in traction batteries results in a disproportionate increase in the number of parts within a battery system due to their relatively small volume compared to prismatic cells and pouch cells. This escalation raises the cost of pack production. Increasing the winding diameter by a factor of $n$ leads to a corresponding increase in the required electrode length, thereby extending the distance that the electric current must travel by approximately $n^2$. This disproportionately long electrode path results in significant inhomogeneities, adversely affecting safety, lifespan, and performance. To mitigate these issues, the electrical and thermal paths within the cell must be designed independently of the winding diameter and arranged axially to facilitate axial current and heat flow. Various solutions have emerged in the literature over the past three years, with some already implemented industrially, each differing in design features and production technology. This chapter contributes to the comprehensive understanding of cylindrical battery cells through a detailed product analysis, leading to a generic product description.

The use of cylindrical battery cells is gaining significance in the automotive industry, exemplified by BMW's new "New Class" platform. BMW has announced a shift away from prismatic cells to cylindrical cells, specifically type 46*xxx* [14]. These cylindrical cells are adjustable in height based on the vehicle model, offering flexibility in design. This paradigm shift is propelled by innovative cell designs that facilitate the expansion of cell volume.

The use of cylindrical battery cells in the automotive sector was pioneered and popularized by Tesla. Initially, in 2006, the company adopted a pragmatic approach by utilizing type 18650 Li-ion cells (18 mm in diameter and 65 mm in height), which have been commercially available since 1994 [15]. Tesla interconnected over 5000 of these cells to achieve the necessary energy capacity for their battery systems. Ten years after launching their first production vehicle, Tesla introduced the 21700 cell format, which is 47% larger (21 mm in diameter and 21 mm in height), aiming to reduce the number of cells and simplify battery design. Both the 18650 and 21700 cells share similar designs and production technologies. The 21700 cells gained prominence through their application in Tesla models 3 and Y.

In 2020, Tesla introduced the larger 4680 cells (46 mm in diameter and 80 mm in height), which are approximately 5.5 times larger than the 21700 cells [16]. This scaling was made possible by innovative design adaptations that facilitate short paths for current and heat dissipation [17]. Achieving this required the adoption of new production technologies, such as laser notching and bending of foil tabs, as well as optimizing the contact between the jelly roll and the cell housing. The production of the 4680 cells simplified some steps compared to the 18650 and 21700 cells.

Notably, intermittent coating and welding of tabs have been eliminated, allowing for a more continuous production process where coating occurs without interruptions and electrode webs are not halted for ultrasonic welding to attach tabs [18].

### 4.3.2 *Product analysis methodology*

To analyze the properties and design features of cylindrical battery cells, a total of 19 cells were examined, divided into four different cell formats: 18650, 21700, 20700, and 4680. These cells were sourced from five renowned manufacturers: Sony/Murata, Panasonic/Sanyo, LG, Samsung SDI, and Tesla. Except for the 4680 battery cell, which was directly sourced from a Tesla vehicle and manufactured by Tesla itself, all other cells were newly procured. Detailed information regarding the specific cells and their nominal capacities is provided in Table 4.1.

#### 4.3.2.1 *Analysis of cell properties*

The cell properties investigated in this study include (i) the energy density, (ii) cell impedance, and (iii) surface and internal cell temperatures. Conventional cell holders were used for contacting the 18650, 20700, and 21700 cells. In contrast, the 4680 cell was specially prepared for contacting, as depicted in Figure 4.3. The contact points on the top were attached in a manner similar to the integration of cells in Tesla battery packs.

#### (1) **Energy density**
The discharge energy was considered in determining the energy density. All cells underwent charging in a climate chamber at an ambient temperature of $T_{amb} = 22°C$ using a constant current and constant voltage (CCCV) method. A charging rate of 0.2 C was applied until reaching a voltage of $U = 4.2$ V. Discharge occurred at a C rate of 1 C up to a voltage of $U = 2.5$ V. During the constant-voltage phase of the charging, the current was reduced to 0.05 C, while during discharge, it was reduced to 0.1 C. Utilizing manufacturer's data for comparing energy density was deemed unsuitable for this investigation, as it is often unclear how these values were obtained.

In cases where the cut-off voltage exceeded 2.5 V – applicable to three cells – the discharge was nonetheless limited to 2.5 V for the purpose

*Energy Density versus Power Density, Lifespan, Safety, and Costs* 101

**Table 4.1.** Investigated battery cells. As the maximum C-rates for the investigated cells differ significantly, the cells are divided into two groups: Group 1 and Group 2.

| Manufacturer | Cell | Nominal cell capacity $C_N$ / mAh |
|---|---|---|
| Sony/Murata | US18650VTC4 (VT4)[†] | 2100 |
| | US18650VTC6 (VTC6)[†] | 3120 |
| | US21700VTC6A (VTC6A)[†] | 4100 |
| Panasonic/Sanyo | UR18650RX (18650RX)[†] | 1950 |
| | NCR18650B (18650B)[*] | 3250 |
| | NCR2070C (2070C)[†] | 3500 |
| | NCR2070B (20700B)[*] | 4000 |
| LG | INR18650-MJ1 (MJ1)[*] | 3500 |
| | INR18650-HG2L (HG2L)[†] | 3000 |
| | ICR18650-HE2 (HE2)[†] | 2500 |
| | INR21700-M50T (M50T)[*] | 4850 |
| | INR21700-M50LT (M50LT)[*] | 4800 |
| Samsung SDI | INR18650-25R (25R)[†] | 2500 |
| | INR18650-35E (35E)[*] | 3400 |
| | ICR18650-22P (22P)[*] | 2150 |
| | INR21700-40T (40T)[†] | 4000 |
| | INR21700-33J (33J)[*] | 3200 |
| | INR21700-50E (50E)[*] | 4900 |
| Tesla | Tesla 4680 | — |

*Note*: [*]First Group; [†]Second Group.

of determining energy density. For all other tests, the cut-off voltages specified in the datasheets were used. To ensure safety during the investigation of the 4680 cell, the current $I$ was limited to $I = 10$ A during both the charging and discharging processes. This precautionary measure was taken due to limited data on the cell's optimal operating conditions. The cell was charged with a current of $I = 10$ A using the CCCV method. During the constant-voltage phase of charging, the current was reduced to $I = 1$ A. After a rest period of two hours, the cell was discharged at the same current, with the current during the constant-voltage phase of discharging reduced to 2 A.

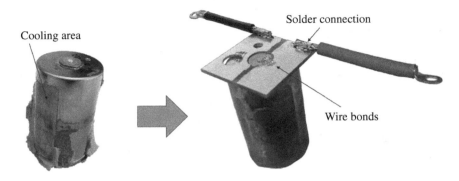

**Figure 4.3.** Contacting of the 4680 battery cell. The cell is contacted on one side, similar to the integration method used in Tesla packs. The single-sided contacting simplifies pack assembly since only one side needs accessibility.

(2) **Impedance**
Electrochemical impedance spectroscopy (EIS) is a well-established technique for analyzing Li-ion battery cells. It quantifies the complex impedance Z, defined as the ratio of voltage $u(t)$ to current $i(t)$, as outlined in equation (4.3). This relationship captures the signal behavior, incorporating the angular frequency $\omega$ and the phase shift $\varphi$. EIS can be conducted using two distinct approaches, both of which provide insights into the electrochemical properties of battery cells by examining their response to electrical excitation:

(a) The galvanostatic method, which is current-driven, applies a current $i(t)$ as the input signal to the electrochemical system under investigation. The system's response, specifically the voltage $u(t)$, is then recorded.
(b) The potentiostatic method, on the other hand, involves stimulating the system with a voltage $u(t)$ and measuring the resultant current $i(t)$:

$$Z = \frac{u(t)}{i(t)} = \frac{U \cdot \sin(\omega t)}{I \cdot \cos(\omega t)}. \tag{4.3}$$

The cells under study were all analyzed using EIS across a frequency spectrum ranging from 0.01 Hz to 200 kHz. These EIS assessments were conducted within a climate chamber, maintained at an ambient temperature of 22°C. Prior to the EIS evaluation, each cell was first discharged and subsequently charged to its nominal voltage using the CCCV protocol to reduce relaxation effects. After charging, a rest period of one hour was applied to minimize temperature effects on the EIS measurements.

## (3) Temperature

Temperature plays a pivotal role in the performance, longevity, and safety of Li-ion batteries. Lower temperatures can diminish performance and lead to lithium deposition during charging. Conversely, elevated temperatures can accelerate battery degradation and expedite the breakdown of the solid–electrolyte interface (SEI) layer and the electrolyte itself. Such conditions can culminate in the thermal runaway of the battery cells. While the surface temperatures of Li-ion cells are commonly monitored, it is widely acknowledged that there is a discrepancy between surface and internal temperatures due to the low thermal conductivity of the electrodes and the separator. The extended thermal pathways inherent in cylindrical cells exacerbate this temperature gradient. In light of this, the current study measures temperatures both at the surface and within the interior of the battery cells. First, an outline of the load profile is presented, followed by a detailed explanation of the measurement methodology used to record temperatures at both the surface and in the winding core.

## (a) *Load profile*

Charging and discharging are conducted using the CCCV profile. During charging, the current in the constant-voltage phase is reduced to a C-rate of 0.05 C. During discharging, the current is reduced to 0.1 C. The voltage range for each cell is taken from the corresponding datasheet. After each charging or discharging process, a passive cooling phase of one hour is implemented. When discharging with $I = 2$ C, the passive cooling is extended to two hours. Further details on the applied load profile are listed in Table 4.2. Due to significant differences in the maximum C-rates of the cells under investigation, the cells are divided into two groups (see Table 4.1). The load profiles for the two groups differ only in the C-rates used during the charging process.

- *Surface temperature*

  Three negative temperature coefficient (NTC) sensors were placed on the cell can to investigate the surface temperature: at the bottom, in the middle, and above the groove. The temperatures measured by these sensors were then averaged.

**Table 4.2.** Load profile used for the investigation of cell properties.

| | Step | Parameter | Termination condition |
|---|---|---|---|
| 1 | Relaxation | $I = 0$ A | $t = 1$ h |
| 2 | CC charging | $C_{char} = 0.5$ C* or $C_{char} = 0.3$ C† | $U = U_{max}$ |
| 3 | CV charging | $U = U_{max}$ | $I \leq I_{(1/20) C}$ |
| 4 | Relaxation | $I = 0$ A | $t = 1$ h |
| 5 | CC discharging | $C_{dischar} = 0.5$ C | $U = U_{min}$ |
| 6 | CV discharging | $U = U_{min}$ | $I \geq -I_{(1/10) C}$ |
| 7 | Relaxation | $I = 0$ A | $t = 1$ h |
| 8 | CC charging | $C_{char} = 0.5$ C* or $C_{char} = 0.3$ C† | $U = U_{max}$ |
| 9 | CV charging | $U = U_{max}$ | $I \leq I_{(1/20) C}$ |
| 10 | Relaxation | $I = 0$ A | $t = 1$ h |
| 11 | CC discharging | $C_{dischar} = 2$ C | $U = U_{min}$ |
| 12 | CV discharging | $U = U_{min}$ | $I \geq -I_{(1/10) C}$ |
| 13 | Relaxation | $I = 0$ A | $t = 2$ h |
| 14 | CC charging | $C_{char} = 0.25$ C* or $C_{char} = 0.2$ C† | $U = U_{max}$ |
| 15 | CV charging | $U = U_{max}$ | $I \leq I_{(1/20) C}$ |
| 16 | Relaxation | $I = 0$ A | $t = 1$ h |
| 17 | CC discharging | $C_{dischar} = 1$ C | $U = U_{min}$ |
| 18 | CV discharging | $U = U_{min}$ | $I \geq -I_{(1/10) C}$ |
| 19 | Relaxation | $I = 0$ A | $t = 1$ h |
| 20 | CC charging | $C_{char} = 1.25$ C* or $C_{char} = 0.5$ C† | $U = U_{max}$ |
| 21 | CV charging | $U = U_{max}$ | $I \leq I_{(1/20) C}$ |

*Note*: *C-rate for charging the battery cells from the first group 2 second group.
†C-rate for charging the battery cells from the first group 2 second group.

- *Internal temperature*
  To examine the internal temperature of the cells under study, NTC sensors (SC30F103V-Amphenol) were installed. Two sensors were embedded in the center of each cell's jelly roll. The cells were drilled at the center of the negative terminal using a 1 mm drill bit within a glovebox filled with an argon atmosphere and then sealed with a two-component epoxy resin adhesive. The sensors were calibrated over a temperature range from 20°C to 60°C.

Drilling into the cell casing may lead to an increase in the cells' impedance due to reduced contact area and potential electrolyte losses. Elevated electrical resistance can result at higher temperatures. Consequently, for assessing internal temperature, only cells exhibiting less than a 10% increase in ohmic resistance relative to their undrilled state were considered. Initially, the cells were charged to their nominal voltage, followed by an EIS measurement. Subsequently, the cells were discharged to the cut-off voltage to facilitate sensor integration within a glovebox under an argon atmosphere. After integration, the cells were recharged to the nominal voltage for a subsequent EIS measurement under identical conditions. This method enabled the detection of significant impedance increases.

EIS measurements enable the detection of only short-term changes in cells immediately following sensor integration. To exclude the possibility of long-term alterations due to sensor integration, we compared the surface temperature of the prepared cells with that of reference cells. Any cell exhibiting a surface temperature difference greater than 10% was deemed unsuitable for evaluating internal temperature and therefore excluded from the assessment.

### 4.3.2.2 Analysis of cell design

Initially, the battery cells were non-destructively examined using computed tomography (CT), allowing for the capture of various cell characteristics. These characteristics include the configuration of the tabs, geometric parameters, and the overhang between the anode and cathode, also known as the anode overhang. This examination also facilitated an analysis of the quality of the winding process. Wavelength analysis, involving the evaluation of grayscale values along a cross-sectional line, enabled the determination of the spacing between all windings and their respective positions. Following this, all cells were opened in a glovebox under an argon atmosphere, and their design features were systematically documented. Prior to the disassembly experiments, the cells were discharged using the CCCV method at a current of 10 A (equivalent to 0.01 C in the constant-voltage phase).

The cells were opened as follows:

1. insulation of the poles,
2. cutting the cover below the groove with a Dremel,

3. cutting through the tabs,
4. removal of the cover and the upper insulating plates,
5. cutting the bottom of the cell,
6. cutting the anode tabs,
7. removal of the lower insulating plates,
8. cutting a slit in the casing parallel to the winding axis (only if the winding cannot be easily pushed out),
9. removal of the jelly roll,
10. removal of the adhesive tape,
11. unwinding and separating the cell components,
12. measuring the geometry and recording the design features.

The described procedure cannot be applied to the 4680 battery cell. Instead, the cell was cut open on both sides – each 12 mm away from the bottom and the top. Subsequently, the remaining casing was removed with a cut parallel to the cell axis. The steps for opening the 4680 battery cells are illustrated in Figure 4.4.

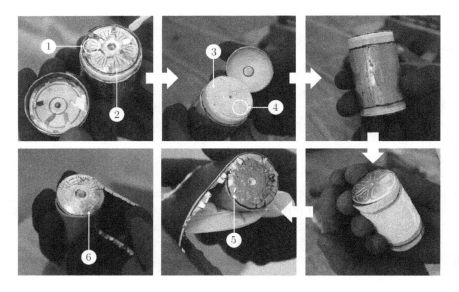

**Figure 4.4.** Opening of the 4680 cell: (1) welding zone, (2) anode tab, (3) cathode tab, (4) welding zone, and (5) notched current collector of the anode, (6) notched current collector of the cathode.

### 4.3.3 Product analysis results

In this section, the findings from the product analysis are presented. The discussion begins with an examination of cell properties. Subsequently, the focus shifts to cell design, with particular emphasis on the jelly roll design and its connection to the cell housing.

#### 4.3.3.1 Cell properties

This section describes the results for cell properties, including energy density, impedance, and the internal and surface temperatures. These properties are compared across all cells.

**(1) Energy density**

Figure 4.5 displays the calculated energy densities for the 19 cells under examination. The gravimetric energy density ranges from 161 to 254.5 Wh/kg, while the volumetric energy density spans from 436.25 to 712.5 Wh/l. There is a linear correlation between gravimetric and volumetric energy densities ($R^2 = 0.97$). It is notable that 21700 cells exhibit higher energy densities on average (264.5 Wh/kg and 629.4 Wh/l)

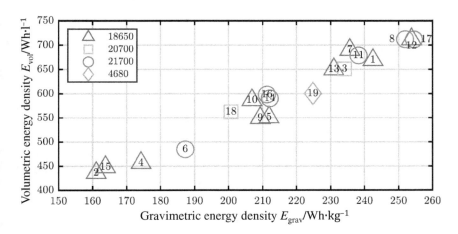

**Figure 4.5.** Gravimetric and volumetric energy density of the tested battery cells: (1) 18650B, (2) 18650RX, (3) 2070B, (4) 22P, (5) 25R, (6) 33J, (7) 35E, (8) 50E, (9) HE2, (10) HG2L, (11) M50T, (12) MJ1, (13) VTC6, (14) VTC6A, (15) VTC4, (16) 40T, (17) M50LT, (18) 2070C, and (19) Tesla 4680.

compared to 18650 cells (192.5 Wh/kg and 575.2 Wh/l). In contrast, the 4680 battery cell demonstrates a relatively lower energy density at 224.8 Wh/kg, despite its electrodes having significantly thicker coatings than other cells. Specifically, the 4680 cell features an anode thickness of approximately 258 $\mu$m and a cathode thickness of 170 $\mu$m, averaged from 10 measurements at multiple electrode positions. In comparison, the average thickness for all other cells is 144 $\mu$m for the anode and 128 $\mu$m for the cathode. Measured energy densities may deviate from the manufacturer's claims due to differences in standardized test conditions. For example, the NCR2070C cell datasheet indicates an energy density of 569 Wh/l and 214 Wh/kg, whereas measurements in this study recorded 562.52 Wh/l and 200.72 Wh/kg, reflecting deviations of 1.1% and 6.2%, respectively.

### (2) **Impedance**

Generally, impedance decreases with an increasing form factor, a trend primarily attributed to the cathode surface area. In Ref. [19], Quinn *et al.* demonstrated that impedance inversely correlates with cathode area. Consequently, a direct comparison among all cells studied is not straightforward. Figure 4.6(a) illustrates the impedance of all tested 18650 cells. When comparing tab designs, two distinct groups emerge: The first group, represented by dashed lines, features cells with two tabs on the anode side and shows low resistance. Four out of five cells exhibit similar ohmic resistance, ranging between 11.5 and 11.7 $\mu\Omega$. However, one cell deviates slightly with a higher resistance of 15.7 $\mu\Omega$. Cells with a single-tab design exhibit ohmic resistance ranging from 14.7 to 36.7 $\mu\Omega$. The 21700 format battery cells exhibit similar behavior (refer to Figure 4.6(b)). Two cells with multi-tab designs demonstrate low ohmic resistance. The VTC6A cell, featuring multi-tab designs on both the anode and cathode sides, achieves minimal resistance at 9.9 $\mu\Omega$. Among cells with traditional tab designs, the Panasonic 20700C cell exhibits the lowest ohmic resistance (see Figure 4.6(c)), utilizing multi-tab designs at both terminals. Detailed images and descriptions of tab designs can be found in Ref. [20] and are elaborated upon in Section 4.3.3.2.

The tabless 4680 battery cell exhibits the lowest impedance, with an ohmic resistance of 6.3 $\mu\Omega$, as shown in Figure 4.6(d)).

### (3) **Temperature**

#### (a) *Surface temperature*

Figures 4.7 and 4.8 display the surface temperatures of the two groups under various C charging and discharging rates. Differences in the

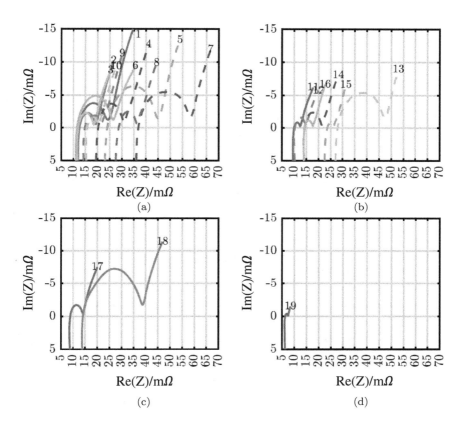

**Figure 4.6.** Nyquist diagrams of the EIS measurements for the battery cells investigated. (a) 18650 cells: (1) VTC4, (2) VTC6, (3) 25R, (4) 35E, (5) 22P, (6) 1865RX, (7) B18650, (8) MJ1, (9) HG2L, and (10) HE2. The dashed curves show the impedance of cells with two tabs on the anode side. (b) 21700 cells: (11) VTC6A, (12) 40T, (13) 33J, (14) 50E, (15) M50T, and (16) M50LT. The dashed curves show the impedance of cells with two tabs on the anode. The VTC6A also has two tabs on the cathode side, which results in a very low impedance. (c) 20700 cells: (17) 20700C and (18) 20700B. Both cells have two tabs on the anode. The 20700C cell also has two tabs at the cathode. (d) 4680 cell: (19) Tesla 4680, first generation: tabless design.

warming behavior of the tested battery cells become evident at a discharge rate of 2 C. The temperature rise ranges from 5 to 24°C for the 18650 cells and from 10 to 24°C for the 21700 cells.

In the following discussion, we exclusively consider the 0.5 C discharge to perform a comparative analysis between conventionally designed cells and the 4680 Tesla design. Charge and discharge tests on

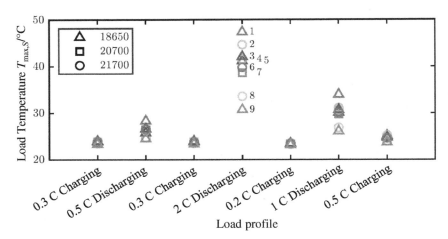

**Figure 4.7.** Mean value of the surface temperature of the first group: (1) 18650B, (2) M50T, (3) MJ1, (4) 50E, (5) 35E, (6) M50LT, (7) 2070B, (8) 33j, and (9) 22P.

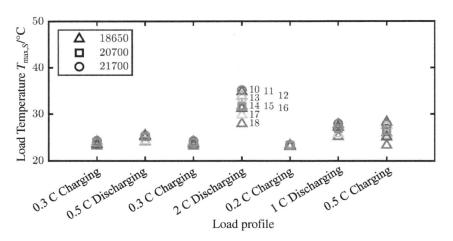

**Figure 4.8.** Mean value of the surface temperature of the second group: (10) 40T, (11) HG2L, (12) VTC6, (13) VTC6A, (14) 2070C, (15) 25R, (16) VTC4, (17) HE2, and (18) 18650RX.

the 4680 cells were conducted using a CCCV profile, as depicted in Figure 4.9. A discharge capacity of 22.2 Ah was achieved, indicating that a current of 10 A corresponds to a discharge rate of 0.5 C. The temperature profiles from the three sensors placed on the cell's surface are illustrated in Figure 4.9.

*Energy Density versus Power Density, Lifespan, Safety, and Costs* 111

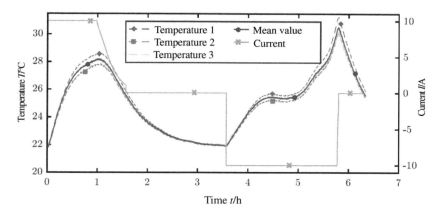

**Figure 4.9.** Temperature on the surface of the 4680 battery cell.

At room temperature in the climate chamber, the external surfaces of the 18650 and 21700 cells with a classic tab design exhibited temperature increases ranging from 1.2 to 5.3°C. In contrast, the 4680 cell experienced a temperature increase of 8.2°C, suggesting that the tabless design does not adequately compensate for the unfavorable surface-to-volume ratio in larger cell formats. The 4680 cells possess only half the surface-to-volume ratio of the 18650 cells. Moreover, over 25% of the electrodes are neither contacted nor structured, which results in longer pathways for heat and electrical current transport.

(b) *Internal temperature*
Only five out of the 18 prepared cells met the criteria for evaluating internal temperature. These criteria, defined in Section 4.3.2, were applied as exclusion criteria: These cells did not exhibit increased electrical resistance after the installation of the temperature sensors, and their surface temperature was comparable to that of the reference cells. At a discharge current of 2C, internal temperature was found to be 30–40% higher than the surface temperature. Table 4.3 lists both the surface and internal temperatures.

Figure 4.10 illustrates the linear correlation between internal temperature and anode thickness, with an $R^2$ value of 0.97, as also confirmed in Ref. [19]. However, this relationship does not hold when considering gravimetric energy density, where the $R^2$ value is 0.7.

Table 4.3. Maximum temperature on the surface of the battery cells and inside the jelly rolls.

| Battery cell | Surface temperature $T_{max,O}/°C$ | Internal temperature $T_{max,I}/°C$ |
|---|---|---|
| 25R | 31.7 | 40.9 |
| VTC6 | 33.9 | 47.3 |
| 35E | 41.5 | 58.3 |
| MJ1 | 42.1 | 57.6 |
| 18650B | 47.4 | 62.5 |

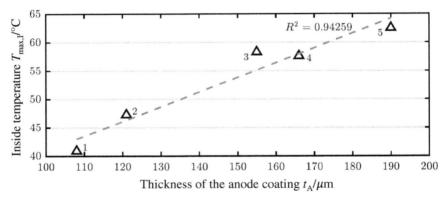

Figure 4.10. Correlation between the temperature in the jelly roll core and anode thickness: (1) 25R, (2) VTC6, (3) 35E, (4) MJ1, and (5) 18650B.

### 4.3.3.2 *Jelly roll design*

This section presents the design features of cylindrical battery cells. Initially, a method to examine the geometry of the jelly roll, referred to as wavelength analysis in this study, is introduced using a selected cell. Following this, the tab design is thoroughly discussed as a focal point of this work. The tab design plays a crucial role in defining the electrical and thermal connections between the jelly roll and the cell can.

### (1) Geometry

The winding geometry is examined by measuring thickness using CT analyses in the wound state. These measurements are subsequently

Energy Density versus Power Density, Lifespan, Safety, and Costs    113

verified in the unwound state after the cells are opened, and they are checked against analytical calculations based on Archimedean spiral.

Figure 4.11 illustrates a representative wavelength analysis using the battery cell INR21700-M50LT. The analysis involved evaluating grayscale values along a marked line perpendicular to an anode tab and a cathode tab. The distances between the peaks of the grayscale values were measured to determine the distances between successive anode and cathode windings. This approach allows for the non-destructive determination of the composite thickness ($t_V$) within the winding using two methods: (i) evaluating anode-to-anode distances and (ii) evaluating cathode-to-cathode distances. The composite thickness is defined by equation (4.4), where $t_S$ represents the separator thickness, $t_C$ denotes the thickness of the cathode coating, $t_A$ denotes the thickness of the anode coating, $t_{Al}$ represents the thickness of the aluminum current collector, and $t_{Cu}$ denotes the thickness of the copper current collector:

$$t_V = 2 \cdot t_S + 2 \cdot t_C + 2 \cdot t_A + t_{Al} + t_{Cu}. \qquad (4.4)$$

The wavelength evaluation was conducted on two sides of the battery cell: the tab-free (homogeneous) side and the side with two tabs (inhomogeneous side). On the homogeneous side, the mean composite thickness and standard deviation were found to be 325.23 ± 32.8 μm (anode-to-anode) and 326.37 ± 25.15 μm (cathode-to-cathode). Conversely, on the

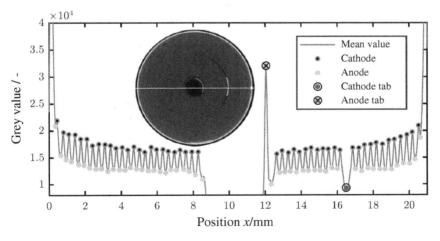

**Figure 4.11.**    CT analysis of the battery cell INR21700-M50LT.

inhomogeneous side, the mean composite thickness and standard deviation were measured as 326.34 ± 25.24 μm and 326.33 ± 25.22 μm, respectively (see Figure 4.12).

The average thickness values for both sides are very similar, resulting in a total average thickness of 326 μm. The standard deviations exceed the typical coating tolerances of the electrodes, which usually range between ±2 and ±5 μm. These differences can be attributed to three factors: (i) the influence of tabs, which compress the coating and cause visible mechanical damage in several consecutive windings; (ii) measurement errors arising from the resolution of the CT scan; (iii) unknown coating tolerances specific to the cells under study.

Subsequently, the validity of the wavelength analysis measurement method was verified through measurements in the open state and confirmed analytically.

After determining the composite thickness ($t_V$) in the wound state via CT scan evaluations, the cell was opened, and the individual components of the composite were measured separately. The following average values were obtained by recording five measurements at randomly selected positions: $t_S = 11$ μm, $t_C = 52.5$ μm, $t_A = 82.5$ μm, another $t_A = 25$ μm, and $t_{Cu} = 10$ μm. This results in a total composite thickness of 327 μm. The discrepancy between the CT and post-mortem measurements is thus less than 0.3%. This consistency indicates that the radius in the wound state

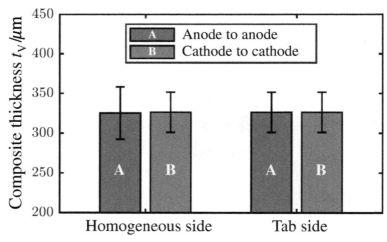

**Figure 4.12.** Evaluation of the composite thickness in the wound state through CT analyses.

*Energy Density versus Power Density, Lifespan, Safety, and Costs* 115

increases precisely by the composite thickness ($t_v$) from one winding to the next (from winding $n$ to winding $n + 1$). Consequently, it can be deduced that the jelly roll is very accurately represented by the Archimedean spiral.

Along the marked line in the CT image (refer to Figure 4.11), 24.5 windings were identified, comprising the entire composite (1 × double-sided coated anode, 1 × double-sided coated cathode, and two separators). Using the corresponding inner and outer diameters of 4 and 19.9 mm, as measured in the CT image, and applying the analytical Archimedean spiral equation (refer to equations (4.5)–(4.8)), 24.4 windings were calculated analytically. The difference between the measured and analytically verified windings is less than 0.4%:

$$t_V = \frac{a}{2}\left[\varphi\sqrt{1+\varphi^2} + \ln\left(\varphi + \sqrt{1+\varphi^2}\right)\right]_{\varphi_i}^{\varphi_a}, \tag{4.5}$$

$$a = \frac{t_V}{2\pi}, \tag{4.6}$$

$$\varphi_i = \frac{r_i + \frac{t_V}{2}}{a}, \tag{4.7}$$

$$\varphi_a = \frac{r_a + \frac{t_V}{2}}{a}. \tag{4.8}$$

$\varphi$ represents the angle of rotation, where $\varphi_i$ and $\varphi_a$ denote the starting and ending points of the Archimedean spiral, respectively. $l_V$ denotes the composite length and consists of two separators: a double-sided coated cathode and a double-sided coated anode. $r_a$ and $r_i$ refer to the outer and inner winding radii, respectively.

## (2) Tab design
In the following, the identified tab configurations are described. There is a distinction between single-tab and multi-tab designs. In a single-tab design, a single tab is utilized on both the anode and cathode sides. In contrast, a multi-tab design employs at least two tabs on either the anode or cathode side, or both. These configurations are denoted as $x$(E or M) $y$(E or M), where $x$ represents the number of anode tabs and $y$ the number of cathode tabs. The letters E and M indicate the tabs' positions on the electrode bands, with E standing for "End" and M for "Middle."

(a) *Single-tab design*
- **Configuration 1E1M:** The anode tab is positioned on the exterior of the winding, while the cathode tab is situated at the center of the electrode (refer to Figure 4.13(a)). Both tabs are attached to the current collectors using ultrasonic welding. The anode tab is folded and welded to the bottom of the housing, whereas the cathode tab is affixed to the cap using laser welding, resistance welding, or, occasionally, ultrasonic welding. Examples of cells employing this configuration include Samsung INR18650 35E, Samsung ICR18650 22P, Panasonic NCR18650B, Samsung INR21700 33J, and LG Chem INR21700 M50T.
- **Configuration 1M1M:** The anode tab and the cathode tab are positioned in the middle of the jelly roll at the same level, as shown in Figure 4.13(b). In this configuration, the overlap between a tab and the electrode should not exceed half the width of the electrode to minimize compressive stress within the jelly roll. The connection techniques are identical to those used in the 1E1M configuration. This design is

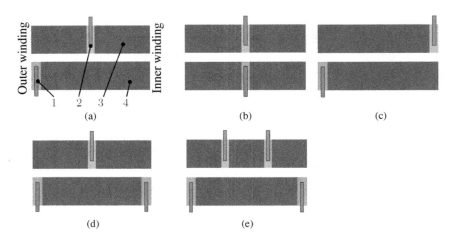

**Figure 4.13.** Tab designs (1 – Ni/Cu-Tab; 2 – Al-Tab; 3 – Cathode; 4 – Anode): (a) tab design in Samsung INR18650 35E, Samsung ICR18650 22P, Panasonic NCR18650B, Samsung INR21700 33J, and LG Chem INR21700 M50T; (b) tab design in Samsung INR21700 40T, and Samsung INR21700 50E; (c) tab design in LG INR18650MJ1; (d) tab design in Sony/Murata US18650VTC4, Sony/Murata US18650VTC6, Samsung INR18650 25R, Sanyo 18650RX, LG INR18650HG2L, LG ICR18650-HE2, Panasonic NCR20700B, and LG INR21700 M50LT; (e) tab design in Sanyo NCR2070C and Sony US21700VTC6A.

employed in the Samsung INR21700 40T and Samsung INR21700 50E battery cells.

- **Configuration 1E1E:** The anode tab is attached to the exterior of the jelly roll, while the cathode tab is situated on the inner diameter, as shown in Figure 4.13(c). This design is employed in the LG INR18650MJ1 cell among the cells that were studied.

(b) *Multi-tab design*

- **Configuration 2E1M:** On the anode side, two tabs are used, attached to the outer and inner diameters at the ends of the electrode (refer to Figure 4.13(d)). On the cathode side, only a single tab is utilized, which is affixed in the middle of the cathode. Example cells with this configuration include Sony/Murata US18650VTC4, Sony/Murata US18650VTC6, Samsung INR18650 25R, Sanyo 18650RX, LG INR-18650HG2L, LG ICR18650-HE2, Panasonic NCR20700B, and LG INR21700 M50LT.
- **Configuration 2E2M:** Both the anode and the cathode have two tabs each. The anode tabs are attached at the ends of the electrode, while the cathode tabs are positioned in the middle (refer to Figure 4.13(e)). This design is utilized in the Sanyo NCR2070C and Sony US21700VTC6A battery cells. A notable feature of these cells is that the cathode tabs are not directly connected to the cap. Initially, the tabs are interconnected. Subsequently, a straightforward connection to the cap is established. The tabs are joined either by ultrasonic welding or by using a tab disc.

(c) *Tesla design*

When the jelly roll is unwound, the laser-notched area of the cathode begins 400 mm from the cathode's starting point and extends to 400 mm before its end. This means 25% of the total cathode length, which is 3180 mm, remains unnotched. Similarly, the structured foil tabs of the anode start at 400 mm and extend to 400 mm before the end of the anode, leaving 24% of the total anode length, which is 3370 mm, unnotched. The notching geometry of the Tesla 4680 battery cell is illustrated in Figure 4.14.

The non-structured areas on the current collectors are intentionally devoid of foil tabs to maintain the necessary flexibility structure in the tab disc's structure. This flexibility enables the separation of connections between the jelly roll and the tab disk as well as between the tab disk and

118  S. Baazouzi & K. P. Birke

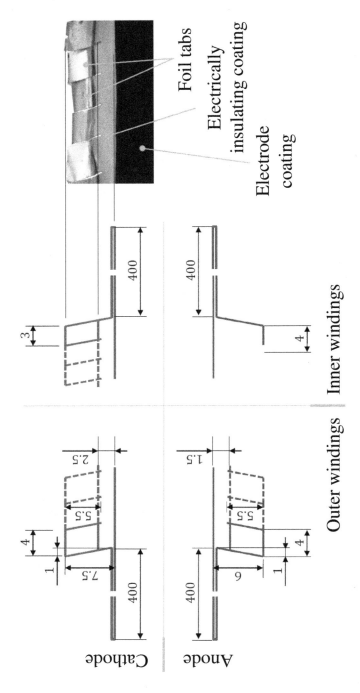

Figure 4.14. Notching geometry of the 4680 Tesla battery cell.

the casing. Such separation is essential to protect the delicate laser-welded seams between the jelly roll and the tab disk from potential damage during both production and usage phases.

During assembly, the tab disks endure significant mechanical stress, particularly during the crimping process that seals the cell. This process involves clamping the tab disk and connecting it to the negative terminal. During operation, the flexible structure of the tab disk can accommodate potential volume expansions and shield the laser-welded seams from mechanical stresses, such as vibrations.

### 4.3.4 *Generic description*

Beyond the design of the electrodes, electrolyte, and separators, two central functions are crucial for the efficient and safe operation of Li-ion battery cells:

(1) The transfer of current and heat between the electrodes and the end faces of the jelly roll.
(2) The transfer of current and heat between the end faces of the jelly roll and the cell housing.

Subsequently, these two functions will serve as a basis for a generic description of the design principles of cylindrical battery cells based on the product analysis conducted (refer to Figure 4.15). The jelly roll design determines the first function, while the second function is defined by the connection between the coil and the cell housing.

### 4.3.4.1 *Jelly roll design*

Fundamentally, two types of jelly rolls can be distinguished: (i) with tabs and (ii) without tabs, also known as "tabless." The tab design can be implemented either as a single-tab design (Design A, Figure 4.15) or as a multi-tab design (Design A'). Tabless design can be achieved either by structuring and folding uncoated areas on the current collectors (Design B) or by using a continuous uncoated area without laser beam structuring (Design C).

120  S. Baazouzi & K. P. Birke

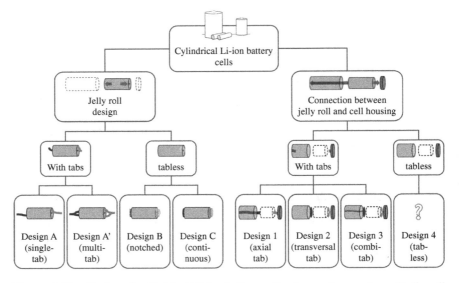

**Figure 4.15.** Design of cylindrical Li-ion battery cells: A generic overview of jelly roll designs and contacting options.

### 4.3.4.2 *Jelly roll contacting*

For connecting the winding with the casing, the following variants are possible:

- The protruding tabs are folded and welded to the bottom and the cover (Design 1, Figure 4.15). Such tabs are referred to as axial tabs in this chapter, as they run parallel to the winding axis.
- A current collector is welded to the end faces of the jelly roll perpendicular to the winding axis. The current collector is then connected to the cell housing (Design 2). Such current collectors are referred to as transversal tabs in this chapter.
- Transversal tabs are extended with strips in a welding or punching process. The strips are then connected to the cell casing (Design 3). Two implementations were identified for this design:
  o Axial tabs are bent around a transversal tab and welded, resulting in a B3 design.
  o A transversal tab is welded to the end face of the jelly roll, leading to a C3 or D3 design.

- The end faces of the winding are directly connected to the cell casing without the use of axial or transversal tabs (Design 4). This design is the subject of current research work.

Figure 4.16 shows exemplary designs of the cells studied:

- **A1 Design (LG M50LT):** A single tab is connected to the cover by laser beam welding with two parallel weld seams.
- **A'1 Design (Murata VTC6A):** Two tabs of different sizes and lengths are initially joined by ultrasonic welding. Subsequently, one tab is connected to the cover by resistance welding.
- **A3 Design (Panasonic 2070C):** Two axial tabs are folded around a transverse tab and connected by laser beam welding using two parallel weld seams each. The transverse tab is then connected to the cover.
- **B2 Design (Tesla 4680, Generation 1):** Partial areas of the current collectors are structured with a laser and bent over at the ends of the jelly roll after winding. Two transverse tabs are attached to the winding using laser beam welding, which are subsequently connected to the housing. CT images of the studied battery cell are displayed in Figure 4.17.

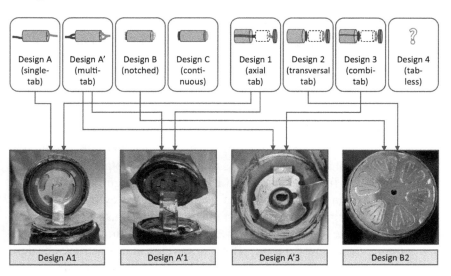

**Figure 4.16.** Example designs for the technical implementation of the connection between the jelly roll and the cell housing.

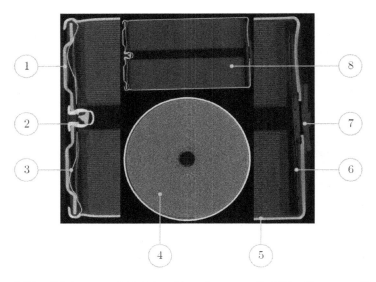

**Figure 4.17.** CT images of Tesla's tabless battery cell 4680: (1) cover, (2) rivet, (3) three-dimensional transversal tab of the anode, (4) cross-section through the battery cell, (5) housing (negative pole), (6) two-dimensional flat transversal tab of the cathode, (7) positive pole, and (8) longitudinal section through the battery cell.

C2 and C3 designs have been documented in the literature [4, 7]. As of the current writing, these designs are not yet available commercially.

### 4.3.5 *Production technologies*

The increasing diversity of products, short product life cycles, and growing demand for cylindrical battery cells necessitate flexible production systems capable of accommodating changing designs and production volumes. Additionally, these systems must exhibit adaptability to respond to unforeseeable future changes and innovations. A significant challenge in this area lies in the production of the jelly roll for cylindrical battery cells, where multiple processes must occur simultaneously. This includes aligning electrode and separator tapes and implementing tab designs. In this section, we present ongoing efforts to develop and construct a design- and format-flexible winding machine capable of accommodating the diverse designs discussed earlier. This machine aims to explore technical production processes for scalable cell designs. Initially, we outline the fundamental requirements for the winding machine, followed by a detailed examination of production steps and specific tasks involved in the winding process. Finally, we discuss the material flow and functional modules of the winding machine.

### 4.3.5.1 *Requirements for the winding machine*

The requirements for the winding system are defined based on transformation enablers such as universality, modularity, scalability, compatibility, and mobility, as outlined in Ref. [21]:

(1) **Universality:** The winding machine is designed and sized to meet diverse requirements in both product design and production technologies. In terms of product design, it must be adaptable to accommodate various design and format requirements. Regarding production methods, it needs to handle a wide range of process parameters and expand process capabilities.
(2) **Modularity:** The winding system is composed of various functional units designed and built modularly for easy replacement as needed.
(3) **Scalability:** The winding machine can be enhanced technically by integrating new functional units, and spatial scaling is feasible if required for technical expansion.
(4) **Compatibility:** The winding machine features standardized interfaces designed for the exchange of information, media, and energy within a production environment.
(5) **Mobility:** The functional units of the winding machine should be segmented into functional groups that can be integrated into various mobile production cells.

### 4.3.5.2 *Winding tasks and production steps*

Jelly rolls for cylindrical Li-ion battery cells feature two fundamental designs: (i) those with tabs and (ii) those without tabs, also known as tabless. Tabless cells can be achieved using either laser-notched foil tabs or a continuous foil tab. The central step in manufacturing jelly rolls is the winding process itself. Compared to stacking, this process is simpler and more continuous, making the production of cylindrical battery cells more cost-effective, robust, and scalable. The winding tasks for the three basic designs are depicted in Figure 4.18: (a) jelly rolls with classic tabs, (b) jelly rolls with notched foil tabs, and (c) jelly rolls with a continuous foil tab.

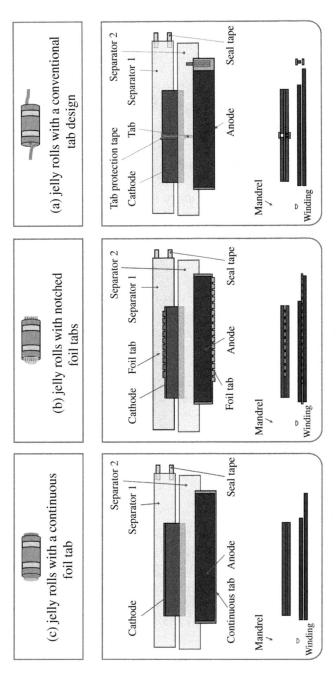

**Figure 4.18.** Schematic representation of the winding tasks in jelly roll production.

## Jelly roll manufacturing

The winding process itself remains consistent across all variations. Prior to winding, appropriate solutions for edge alignment and positioning are applied to adhere to winding manufacturing tolerances, minimize telescoping effects, and prevent short circuits within the cell. The conventional method of winding production (depicted in Figure 4.18(a)) involves the following steps:

(1) Isolating the tabs at the winding exit point. Specifically, only the cathode tabs are isolated, given that the anode is wider than the cathode.
(2) Ultrasonic welding of the tabs on the uncoated areas of the electrodes, requiring intermittent coating.
(3) Applying protective tape to the tabs.
(4) Picking up the separators and starting the winding process without electrodes.
(5) Inserting the anode and continuing the winding process without the cathode.
(6) Inserting the cathode and continuing the winding process with the entire composite consisting of two separators, an anode, and a cathode.
(7) Cutting the cathode strip and continuing to wind the remaining winding components.
(8) Cutting the anode strip and continuing to wind the separators to close the winding.
(9) Cutting the separators.
(10) Securing the winding with one or more sealing tapes or through a heat-sealing process.
(11) Testing. The most relevant measurements include height, diameter, weight, overhang between anode and cathode, and impedance.

Steps 4–1 are also necessary for producing jelly rolls with notched foil tabs (refer to Figure 4.18(b)) and jelly rolls with a continuous foil tab (refer to Figure 4.18(c)). Additional upstream processes are required for jelly rolls with notched foil tabs. These processes include the following:

(1) Laser beam notching of the uncoated areas of the electrode webs. This process presents several challenges, particularly: (i) ensuring precise positioning of the electrode edges; (ii) minimizing material ejection to

126   S. Baazouzi & K. P. Birke

prevent short circuits within the battery cell, achieved by maintaining a safety margin between the end of the cutting edge and the start of the electrode coating; and (iii) reliably removing cut materials and sparks through appropriate extraction solutions.

(2) Prebending of the uncoated foil tabs is performed immediately before the winding unit.

(3) Bending of the laser-notched foil tabs: During this research, no information was available in the literature regarding the technical realization of this process. Therefore, a bending process was developed using a conventional lathe. Prior to the winding process, the foil tabs are prebent at angles ranging from 30° to 45°. After winding, a single tool moves continuously perpendicular to the rotating jelly roll, incrementally bending the foil tabs to an angle of up to 90°. Key process parameters include the rotational speed of the jelly roll holder, the feed rate, and the surface condition of the bending tool.

## Quality parameters

Commercial Li-ion cells with negative graphite electrodes typically exhibit an anode overhang, where the negative electrode is slightly wider than the positive electrode. In the examined cells, the anode overhang ranges from 0.5 to 1 mm. This design ensures that each cathode has a corresponding anode, even if the electrodes are not perfectly aligned during the winding process. If a section of the cathode lacks a matching negative electrode, lithium ions and electrons produced in this area can deposit as lithium metal on the nearest part of the current collector, typically at the closest edge of the negative electrode. This scenario can pose safety risks, which is why nearly all Li-ion cells with graphite negative electrodes are equipped with an anode overhang [22]. The following geometric dimensions are critical quality parameters in the manufacturing of jelly rolls: the highest anode to the lowest cathode $(P_1)$, the lowest anode to the highest cathode $(P_2)$, the lowest cathode to the highest cathode $(P_3)$, the maximum overhang from anode to cathode $(P_4)$, and the minimum overhang from anode to cathode $(P_5)$. The value ranges for the cells investigated in this study are presented in Table 4.4.

### 4.3.5.3 Material flow and modules of the winding machine

A modular solution consisting of 14 functional units was developed and implemented in a winding machine (see Figure 4.19).

Table 4.4. Geometric parameters on the end faces of the jelly rolls.

| Parameter | Negative terminal (mm) | Positive terminal (mm) |
|---|---|---|
| $P_1$ | 0.2–0.86 | 0.15–0.7 |
| $P_2$ | 0.1–0.6 | 0.2–0.45 |
| $P_3$ | 0.1–0.6 | 0.1–0.45 |
| $P_4$ | 0.45–1.15 | 0.4–1.05 |
| $P_5$ | 0.3–0.7 | 0.25–0.9 |

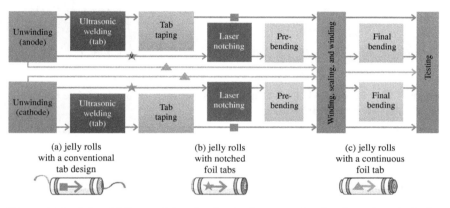

**Figure 4.19.** Material flow and functional units for processing the electrode webs in the production of different jelly roll designs: (a) conventional tab design, (b) notched foil tabs, and (c) continuous foil tab. Functional units for processing the two separators, positioning and aligning electrode webs, and adjusting web tension are not depicted, as they are identical for all jelly roll designs.

*Source*: Images of the jelly rolls: proprietary designs graphically rendered by Michael Fuchs, Fraunhofer IPA.

The main systems and assemblies of the machine are depicted in Figure 4.20, including the following:

(1) A human–machine interface (HMI) for recipe management and system operation and configuration.
(2) Unwinding systems for electrodes with integrated web tension control.
(3) Splicing devices for electrodes.
(4) Edge control systems for positioning electrode strips before ultrasonic welding of tabs.
(5) Robots for flexible tab handling.

128  S. Baazouzi & K. P. Birke

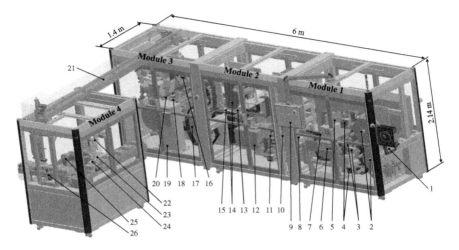

**Figure 4.20.** Realized winding machine for the design- and format-flexible production of jelly rolls for cylindrical Li-ion battery cells.

*Source*: The CAD model of the machine was created by Matthias Strecker (acp systems AG).

(6) Tab magazines for storing tabs of different geometries.
(7) Two ultrasonic welding units with dual welding zone geometries. Multiple welding zones can be positioned flexibly.
(8) Clamping device 1 for threading the electrode webs.
(9) A HMI for configuring parameter sets for the ultrasonic welding process.
(10) Taping station for applying protective adhesive tapes to the ultrasonically welded tabs.
(11) Edge control systems for positioning the electrode webs before laser notching.
(12) Notching unit for laser notching of foil tabs for tabless designs, equipped with an integrated spark and cut catcher.
(13) Drive rollers for conveying the electrode webs.
(14) Dancer systems for decoupling discrete and continuous process steps, controlling the tension of the electrode strips immediately before winding. Electrode tension can be dynamically adjusted during the winding process.
(15) Clamping device 2 for threading the electrode strips.
(16) Edge control systems (4×) for positioning the electrode and separator strips before the winding process.
(17) Unwinding systems for the separators with integrated control for web tension and deionization units.

(18) Feeding systems for the electrodes and separators, equipped with cutting units and suction device for removing electrode particles generated during cutting.
(19) Winding unit with different mandrel diameters.
(20) Pre-bending unit for pre-bending foil tabs.
(21) Pneumatic axes for automated workpiece transfer.
(22) Folding unit for bending foil tabs for tabless cell designs.
(23) Camera system for measuring winding height and diameter.
(24) High-pot testing with a flexible winding contact system.
(25) Robot for handling the windings in module 4.
(26) Removal system consisting of conveyor belts and intelligent workpiece carriers for recording ambient conditions.

The winding machine was developed by acp systems AG and consists of two parallel production lines for processing the anode and cathode. Both lines are illustrated schematically in Figure 4.19. The functional units highlighted in green are identical across all winding designs. Other units can be flexibly activated or deactivated depending on the cell design, encompassing basic configurations and facilitating various design combinations. These configurations include: (i) windings with conventional tabs, (ii) windings with structured foil tabs, and (iii) windings with a continuous foil tab, along with all six possible design combinations, which are depicted in Figure 4.21.

### 4.3.5.4 Outlook

Throughout this section, a comprehensive analysis of cylindrical battery cells was conducted, facilitating the development of a comprehensive

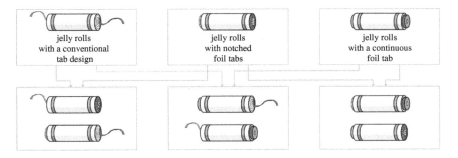

**Figure 4.21.** Overview of the basic designs and design combinations.

*Source*: Images of the Jelly Rolls: original designs, graphically rendered by Michael Fuchs, Fraunhofer IPA.

product description. The insights obtained were pivotal in the development of a modular and adaptable winding machine. Special emphasis was placed on incorporating transformation enablers, which played a crucial role in the conception and design of the machine. This approach allowed for the machine's adaptation to various product requirements and specifications, making it suitable for diverse applications. Three fundamental jelly roll designs were identified: (i) jelly rolls with a conventional tab design, (ii) jelly rolls with notched foil tabs, and (iii) jelly rolls with a continuous foil tab. The latter two designs are commonly referred to as tabless cells in the literature. These designs aim to address the challenge of long current paths, although the electrical path remains relatively long due to utilizing the cell casing as a conductive route. This is primarily because the cell connectors for interconnecting cylindrical battery cells in modern cell-to-pack battery architectures are typically located on one side of the cell. Cells with one-sided contact between the jelly roll and the cell casing in a tabless design offer significant potential to reduce the impact of the cell casing on electrical resistance, shorten assembly time, and increase energy density. Moreover, cylindrical battery cells with one-sided contact provide considerable advantages for recycling-friendly cell designs, facilitating easier cell disassembly and thereby enhancing the quality and purity of recycled materials. The winding machine developed in this study enables the implementation of these innovative one-sided contact designs.

## 4.4 Power Density

In this section, we delve into the aspect of power density in Li-ion battery cells. Section 4.4.1 explores the trade-off between power density and energy density, emphasizing the balance that must be carefully considered. Section 4.4.2 outlines the factors influencing power density, providing insights into the variables at play. Finally, in Section 4.4.3, we present a qualitative analysis of power density across the three cell formats: cylindrical, prismatic, and pouch.

### 4.4.1 *Trade-off between energy density and power density*

In EV applications, the trade-off between energy density and power density in Li-ion battery cells is a critical design consideration that directly impacts vehicle range and charging capabilities. Energy density refers to the amount of energy a battery can store relative to its mass or volume, determining the potential range of an EV on a single charge. Conversely, power density, expressed in W/kg or W/l, reflects the rate at which a battery can deliver or absorb energy, influencing acceleration performance and the feasibility of fast charging. High power density is crucial for enabling fast charging, allowing the battery to accept rapid energy influx without overheating or degrading its lifespan.

However, optimizing for one attribute often comes at the expense of the other due to material and design constraints. Batteries with high energy density typically utilize dense electrode materials and designs that restrict the speed of ion transport, thereby reducing power density. This limitation can result in longer charging times and may not meet the expectations of consumers who expect quick charging comparable to conventional internal combustion engine vehicles. Conversely, cells optimized for high power density may incorporate electrode materials and architectures that facilitate rapid ion exchange but have less volumetric capacity for energy storage, thus reducing the vehicle's range.

The challenge for researchers and engineers lies in balancing these properties through innovative materials science and cell design, with the goal of extending EV range while minimizing charging times. Advances in electrode materials, electrolyte formulations, and cell architecture are crucial to achieving this balance, ensuring that future Li-ion batteries meet the evolving demands of the EV market.

### 4.4.2 *Influencing factors*

This section explores the diverse factors that affect power density, including material selection, cell design, operational conditions, manufacturing quality, and the battery management system.

### 4.4.2.1 *Materials properties*

The choice of electrode materials in Li-ion batteries critically impacts their power density. Typically, the cathode is composed of lithium metal oxides, with variations such as cobalt, manganese, or nickel influencing both voltage and capacity. While materials with higher voltage capacity enhance energy density, they may not necessarily improve power density.

In contrast, the anode, commonly graphite-based, can benefit from alternatives such as silicon or lithium titanate, which offer faster charge and discharge rates, thereby boosting power density. The ionic conductivity of the electrolyte is also crucial; higher conductivity facilitates faster ion transport, increasing power output. The separator plays a pivotal role in preventing short circuits while enabling ion flow. Its material and structure are critical in minimizing internal resistance and maximizing power density.

### 4.4.2.2 Cell design

The design of a Li-ion battery cell, including the thickness and structure of its electrodes, directly impacts its power density. Thicker electrodes can increase energy storage capacity but may also extend the diffusion path for ions, potentially slowing charge and discharge rates and adversely affecting power density. Optimizing electrode porosity is crucial to achieving a balance between ion transport and mechanical strength. Current collectors, typically made of aluminum for the cathode and copper for the anode, must be designed to minimize resistance while remaining robust enough to handle high currents. The overall cell architecture, including the arrangement of these components, is engineered to minimize internal resistance and maximize the efficiency of electron and ion flow, essential for achieving high power density.

### 4.4.2.3 Operational conditions

The operational conditions under which a Li-ion battery is used can significantly affect its power density. Temperature is a critical factor: Both high and low temperatures can increase internal resistance and reduce power output. Batteries tend to perform best within a moderate temperature range. The SoC also affects performance: Lower SoC levels may limit the battery's ability to deliver power compared to when fully charged. Additionally, the rate of charging or discharging (C-rate) influences temperature and internal resistance, thereby affecting power density. Over time, batteries undergo chemical changes and mechanical stress, resulting in decreased power density. Operating batteries outside their optimal temperature range or at high C-rates can accelerate this degradation.

### 4.4.2.4 *Manufacturing quality*

The manufacturing process of Li-ion batteries requires stringent control to ensure high quality and consistency, crucial for achieving optimal power density. Variations in the coating thickness of electrode materials, inconsistencies in electrode composition, or defects in the separator can create areas of high resistance, thereby reducing the overall power density of the cell. Precise alignment and assembly of the cell components are essential to minimize internal resistance. Automated manufacturing processes with strict quality control measures should be employed to ensure that each cell meets the required specifications for optimal performance.

### 4.4.2.5 *Battery management system*

The battery management system (BMS) plays a crucial role in protecting the battery from operating beyond safe parameters. It monitors the battery's state, calculates secondary data, reports this information, controls its environment, and balances its cells. Moreover, the BMS can significantly enhance the power density of a battery by optimizing charge and discharge processes, maintaining operation within the ideal temperature range, and preventing conditions that could increase resistance or cause damage. Additionally, BMS algorithms manage the battery's SoC and health, thereby extending its lifespan and maintaining high power density over time. The BMS is crucial for maximizing battery performance while ensuring safety and longevity.

### 4.4.3 *Power density of cell formats – Qualitative view*

To the best of our knowledge, there is no literature comparing the power density of different cell formats. However, the three primary cell formats exhibit distinct characteristics in terms of power density: (1) Cylindrical cells, such as the 18650 and 21700 formats, are known for their standardized sizes, offering good mechanical stability and excellent thermal behavior due to their geometry. This design facilitates effective heat dissipation, which contributes to higher power density by enabling sustained high rates of power delivery without excessive thermal buildup. However, cylindrical cells have lower space efficiency compared to other form factors due to the unused space when packing them together. (2) Prismatic cells, encased in hard shells typically made of aluminum, are designed to fit into space-efficient rectangular battery packs. While they can be optimized for higher energy density, their power density can be somewhat lower than cylindrical cells. This is partly due to the challenges in managing heat within the larger, flat surfaces of prismatic cells, which can limit the rate of power delivery and dissipation. (3) Pouch cells feature flexible, lightweight, and thin foil-like enclosures, allowing for various sizes and shapes that optimize space utilization within battery packs. Pouch cells can achieve high energy and power densities due to their large electrode surface area and short path for ion transport. However, their power density can be influenced by their ability to dissipate heat and maintain structural integrity during high-power charge and discharge cycles.

In summary, while cylindrical cells may offer higher power density due to their efficient thermal management, pouch cells can also achieve high power densities through design flexibility and electrode optimization. Prismatic cells, though space-efficient, may exhibit slightly lower power density due to their thermal and structural constraints. It's important to note that advancements in materials and cell design continue to blur these distinctions as manufacturers optimize each form factor for specific applications and performance criteria.

## 4.5 Costs, Safety, and Lifespan: Outlook

While energy and power density are critical parameters in the performance of Li-ion battery cells for EVs, it is imperative to consider the broader implications of cost, safety, and lifespan aspects to ensure sustainable and practical energy storage solutions.

Costs are a pivotal factor in the widespread adoption of EVs. The economic viability of Li-ion batteries hinges on reducing production costs through advances in material science, manufacturing processes, and economies of scale. Strategic sourcing of raw materials and innovations in battery design are essential to make EVs competitive with traditional combustion engine vehicles.

Safety is paramount, as the high energy content in Li-ion batteries poses risks such as thermal runaway and fires. Research into BMS, robust cell design, and advanced materials aims to mitigate these risks. Ensuring the safe operation of Li-ion batteries throughout their lifecycle is not only a technical challenge but also a public assurance imperative.

*Energy Density versus Power Density, Lifespan, Safety, and Costs*   135

Lifespan aspects, encompassing circular economy solutions and recycling, are vital for both environmental and economic sustainability. End-of-life management of Li-ion batteries presents opportunities for reuse in secondary applications, such as energy storage systems, before recycling. Recycling technologies are evolving to recover valuable materials such as lithium, cobalt, and nickel, reducing the need for virgin resource extraction and minimizing the environmental footprint.

Integrating circular economy principles into the lifecycle of Li-ion batteries is crucial. This requires a systemic approach that includes design for disassembly, standardized processes for refurbishment, and efficient recycling methods. These measures extend the economic value of the batteries and align with global sustainability goals.

In conclusion, addressing the costs, safety, and lifespan aspects of Li-ion batteries within the framework of circular economy solutions is essential for the responsible growth of the EV sector. Stakeholders must collaborate to innovate and implement practices that will drive the industry toward a more sustainable and economically viable future.

## 4.6 Discussions

In the concluding discussion of this chapter, it is crucial to acknowledge the need for balance among these parameters. Optimizing for power density often necessitates compromises in energy density. Geometric changes, such as reducing layer thickness and increasing layer count, can enhance power density but may reduce energy density by up to 50% when designing a cell for high power output. Despite this trade-off in energy density, these design choices can also contribute to extending the cell's operational life.

Recent advancements in Li-ion battery components, such as separators and electrolytes, appear to have reached a plateau, leaving limited scope for further significant improvements. Consequently, the primary strategy for enhancing power and energy output now revolves around redesigning the battery structure. However, the pursuit of higher energy storage capacities has raised safety concerns. While more energy-dense batteries could potentially reduce the cost per unit of energy, their increased energy content also amplifies the risk of overheating. This can lead to shorter lifespans, safety hazards, and harmful reactions such as lithium plating. Conversely, lower energy densities can have a positive impact on battery lifespan and safety by reducing thermal irregularities within the cells.

Looking ahead, larger cylindrical cells show significant potential, provided that issues related to current collectors can be effectively resolved. Their inherent structural advantages, such as tight winding, could potentially enhance overall performance. Furthermore, these cells are well suited for cell-to-pack battery designs, which have the potential to streamline production processes.

In conclusion, achieving the ideal balance between power and energy densities requires careful consideration of longevity, safety, and economic factors, necessitating a complex interplay of trade-offs. While the choice of geometrical configurations and materials is critical for enhancing power and energy density, it is essential to prioritize safety and durability to achieve reliable and cost-effective battery technology. Looking forward, if technical challenges can be adequately tackled, cylindrical cells may emerge as the preferred choice, paving the way for advancements in battery architecture and performance.

# References

[1] J. Warner, Lithium-ion battery packs for EVs. In: *Lithium-Ion Batteries*. Elsevier, 127–150, 2014.

[2] Semco Infratech. Production Process of Stacked & Wound Batteries: https://www.semcoinfratech.com/stacked-vs-wound-batteries-the-shocking-showdown/#:~:text=Higher%20Battery%20Energy%20Density%20The,in%20comparison%20to%20wound%20batteries. (accessed on 12 March 2025).

[3] A. Fill, M. Kopp, J. Hemmerling, *et al.* Three-dimensional model of a cylindrical Lithium-Ion Cell – influence of cell design on state imbalances and fast-charging capability. In: *2023 IEEE 32nd International Symposium on Industrial Electronics (ISIE)*, 1–8, 2023.

[4] H. Pegel, D. Wycisk, A. Scheible *et al.*, Fast-charging performance and optimal thermal management of large-format full-tab cylindrical lithium-ion cells under varying environmental conditions, *Journal of Power Sources,* 556, 232408, 2023.

[5] S. Link, C. Neef, and T. Wicke. Trends in automotive battery cell design: A statistical analysis of empirical data, *Batteries*, 9(5), 261, 2023.

[6] M. Ank, A. Sommer, K. Abo Gamra *et al.* Lithium-ion cells in automotive applications: Tesla 4680 cylindrical cell teardown and characterization, *Journal of the Electrochemical Society*, 10.1149/1945-7111/ad14d0, 2023.

Energy Density versus Power Density, Lifespan, Safety, and Costs 137

[7] H. Pegel, D. Wycisk, and D. U. Sauer, Influence of cell dimensions and housing material on the energy density and fast-charging performance of tables cylindrical lithium-ion cells, *Energy Storage Materials*, 60, 102796, 2023.

[8] B. Ederer, S. Schillmöller, and Mehr Leistung, $CO_2$-reduzierte Produktion, Kosten deutlich reduziert: Die BMW Group setzt in der Neuen Klasse ab 2025 innovative BMW Batteriezellen im Rundformat ein. https://www.press.bmwgroup.com/deutschland/article/detail/T0403470DE/mehr-leistung-co2-reduzierte-produktion-kosten-deutlich-reduziert:-die-bmw-group-setzt-in-der-neuen-klasse-ab-2025-innovative-bmw-batteriezellen-im-rundformat-ein?language=de (accessed on 12 September 2023).

[9] K.-J. Lee, K. Smith, A. Pesaran, and G.-H. Kim, Three dimensional thermal-, electrical-, and electrochemical-coupled model for cylindrical wound large format lithium-ion batteries. *Journal of Power Sources*, 241, 20–32, 2013.

[10] J. Hemmerling, J. Guhathakurta, F. Dettinger *et al.*, Non-uniform circumferential expansion of cylindrical Li-Ion cells – The potato effect, *Batteries*, 7, 61, 2021.

[11] S. V. Erhard, P. J. Osswald, P. Keil, E. Höffer, M. Haug, A. Noel *et al.* Simulation and measurement of the current density distribution in lithium-ion batteries by a multi-tab cell approach, *Journal of the Electrochemical Society*, 164, A6324–A6333, 2017.

[12] F. Brauchle, F. Grimsmann, O. von Kessel, K. P. Birke, Direct measurement of current distribution in lithium-ion cells by magnetic field imaging, *Journal of Power Sources*, 507, 230292, 2021.

[13] T. Waldmann, R.-G. Scurtu, D. Brändle *et al.*, Increase of cycling stability in pilot-scale 21700 format Li-Ion cells by foil tab design. *Processes* 9, 1908, 2021.

[14] BMW Group. Runde Batteriezellen für die Neue Klasse. https://www.bmwgroup.com/de/news/allgemein/2022/gen6.html (accessed on 21 November 2023).

[15] Panasonic Group. Innovative Product: Lithium ion rechargeable battery. https://holdings.panasonic/global/corporate/about/history/chronicle/1994.html (accessed on 15 September 2023).

[16] S. Baazouzi, N. Feistel, J. Wanner *et al.*, Design, properties, and manufacturing of cylindrical Li-ion battery cells – A generic overview, *Batteries*, 9, 309, 2023.

[17] K. Tsuruta, M. E. Dermer, and R. Dhiman, A cell with a tabless electrode. WO2020096973 (A1), 2020.

[18] Fraunhofer-Institut für System- und Innovationsforschung ISI. Batterie-Update: Potenziale von 46-mm-Rundzellen – Auf dem Weg zum neuen Standardformat. https://www.isi.fraunhofer.de/de/blog/themen/batterie-update/46-mm-rundzellen-potenziale-standardformat-batteriezellen.html (accessed on 8 June 2024).

[19]  J. B. Quinn, T. Waldmann, K. Richter *et al.*, Energy density of cylindrical Li-Ion cells: A comparison of commercial 18650 to the 21700 cells. *Journal of the Electrochemical Society,* 165, A3284–A3291, 2018.

[20]  S. Baazouzi, N. Feistel, J. Wanner *et al.*, Design, properties, and manufacturing of cylindrical Li-ion battery cells – A generic overview: Supplementary materials. 10.5281/zenodo.7798920, 2023.

[21]  P. Nyhuis, G. Reinhart, and E. Abele, (Eds). *Wandlungsfähige Produktionssysteme: Heute die Industrie von morgen gestalten.* PZH Produktionstechnisches Zentrum, Garbsen, 2008.

[22]  B. Gyenes, D. A. Stevens, V. L. Chevrier *et al.*, Understanding anomalous behavior in coulombic efficiency measurements on Li-ion batteries. https://iopscience.iop.org/article/10.1149/2.0191503jes

© 2025 World Scientific Publishing Company
https://doi.org/10.9789811282058_0005

# Chapter 5

# The Role of Raw Materials in Enhancing or Limiting Energy Density

**Daniel Steffen Reichert[*] and Kai Peter Birke[†]**

*Fraunhofer IPA, Nobelstrasse 12, Stuttgart, Germany*

*[*]daniel.steffen.reichert@ipa.fraunhofer.de*

*[†]kai.peter.birke@ipa.fraunhofer.de*

## 5.1 Alternative Battery Technologies

In the pursuit of sustainable and efficient energy storage solutions, research into alternative battery technologies that can complement or even replace the current state-of-the-art lithium-ion battery (LIB) is essential. This chapter aims to identify a viable competitor among emerging battery technologies that can illustrate fundamental design principles and their impact on critical battery parameters. The focus will be on sodium-ion batteries (SIBs) as a potential alternative to conventional LIBs. The discussion will emphasize key factors such as energy density, power density, safety, and cost, which are crucial for understanding the trade-offs and advantages of SIBs. Additionally, the chapter delves into the design principles for electrode materials in SIBs, covering vital concepts such as transport properties, size effects, and the morphology and structure of the utilized materials. This comprehensive approach will provide insights into how these design principles affect the performance and other essential battery parameters, using SIBs as a case study.

### 5.1.1 Need for alternative battery technologies and cell chemistries

The increasing demand for sustainable and cost-effective energy storage solutions has sparked interest in alternative battery technologies. While LIBs currently dominate the market due to their high energy density and performance, several challenges and limitations underscore the need for alternative battery technologies.

Lithium reserves are primarily concentrated in specific geographical regions, particularly the "Lithium Triangle" in South America (Chile, Argentina, and Bolivia), Australia, and China. This concentration leads to geopolitical risks and potential supply chain disruptions. Dependence on a limited number of suppliers can result in significant supply chain disruptions, particularly during periods of political instability or economic sanctions. Figure 5.1 displays the critical and near-critical elements and highlights the commonly used elements in LIBs and SIBs.

The growing demand for lithium, driven by electric vehicles and stationary energy storage, has caused significant price fluctuations. This volatility complicates planning and cost control for LIB manufacturers, leading to higher production costs and increased market instability. Consequently, this makes investments in battery production riskier.

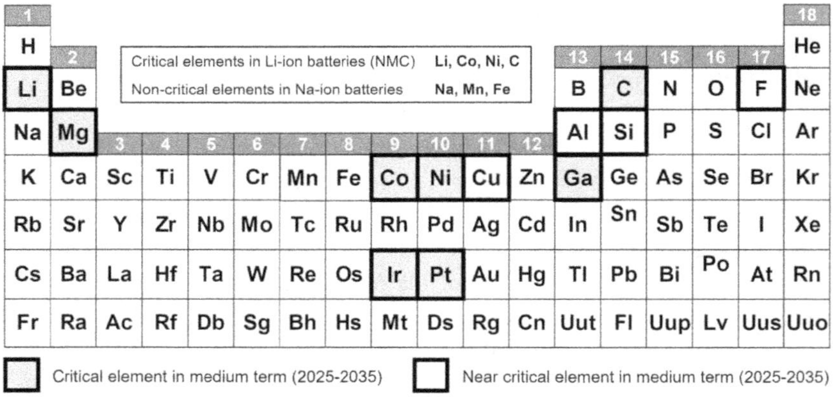

**Figure 5.1.** Critical and near-critical elements for the medium term (2025–2035), as defined by the U.S. Department of Energy (July 2023). Graphite is defined as a critical element among various forms of carbon. Elements commonly used in LIBs and SIBs are also highlighted.

*Source*: Adapted from Ref. [1].

The extraction and processing of lithium are associated with significant environmental concerns, including high water consumption and environmental damage in mining areas. Particularly in saline desert regions, lithium mining can lead to substantial ecological and social issues, including the displacement of local communities and harm to ecosystems [2]. Additionally, the challenges related to the safe disposal and recycling of LIBs exacerbate their environmental impact [3].

LIBs are prone to thermal runaway, which can result in fires and explosions. These safety risks necessitate stringent safety measures and management systems, particularly in large-scale applications such as electric vehicles and stationary energy storage systems. The high energy densities of LIBs exacerbate these risks, as they have the potential to release significant amounts of energy [4].

The materials used in LIBs, such as cobalt, nickel, and manganese, are limited and often concentrated in politically unstable regions. The extraction of cobalt, predominantly in the Democratic Republic of Congo, raises significant ethical and environmental concerns. These materials are not only expensive but also environmentally damaging and difficult to recycle. Moreover, the supply chains for LIBs are complex and susceptible to disruptions. Dependence on a few suppliers and manufacturers can lead to bottlenecks and delays, especially during periods of geopolitical tension or economic uncertainty. This fragility in the supply chain can significantly affect the production and availability of batteries [2, 5].

The production costs for LIBs are high, primarily due to the cost of raw materials such as lithium and other key components. These elevated costs can hinder the widespread adoption of battery technologies in various applications, especially those with a strong focus on total cost of ownership (TCO). Furthermore, the high production costs can lead to longer payback periods for investments in battery production.

### 5.1.2 *Alternatives to lithium-ion batteries*

Given the limitations and challenges associated with LIBs, the search for alternative battery technologies has intensified. Several promising technologies are emerging as viable alternatives, each offering distinct advantages in terms of availability, cost, safety, and performance:

**Sodium-ion batteries (SIBs)** are gaining attention due to the abundant availability and low cost of sodium. SIBs operate similarly to LIBs by shuttling monovalent sodium ions between the anode and cathode. While they currently lag behind LIBs in terms of energy and power density, SIBs present a potentially safer and cost-effective alternative [5].

**Zinc-ion batteries (ZIBs)** hold promise as an environmentally friendly and cost-effective energy storage solution due to their use of abundant and less toxic materials. However, the commercialization of ZIBs encounters challenges, including suboptimal energy density and cycling stability, which are largely attributed to issues with the stability of the currently used anode and cathode materials. Advances in interface engineering are crucial for addressing these issues and developing more stable and efficient materials for ZIBs [6].

**Magnesium-ion batteries (MIBs)** present the potential for higher volumetric capacities due to their divalent nature, carrying two charges per ion. Magnesium's lower reactivity compared to lithium reduces safety risks. However, MIBs face significant challenges that must be addressed before they can be widely adopted. These challenges include the development of suitable electrolytes and cathode materials, as well as improving the overall stability and efficiency of the battery components [7].

**Aluminum-ion batteries (AIBs)** are emerging as a promising low-cost energy storage solution due to aluminum's abundance as the most prevalent metal in the Earth's crust. Aluminum's high volumetric capacity, which is four and seven times larger than that of lithium and sodium, respectively, offers the potential for significant improvements. Although research on rechargeable aluminum batteries dates back to the 1970s, interest has surged since 2010, when the feasibility of an ambient-temperature aluminum system was demonstrated. AIBs potentially offer fast-charging capabilities and enhanced safety, as aluminum's lower reactivity reduces the risk of fires and explosions [8, 9].

**Lithium–sulfur batteries (Li–S)** have attracted significant attention due to their high theoretical energy density, which is several times greater than that of conventional LIBs. The use of sulfur, which is abundant and cost-effective, further enhances their appeal. Despite their potential, Li–S batteries face challenges, such as the "shuttle effect," where soluble polysulfide intermediates migrate between the electrodes, leading to capacity fade and reduced efficiency. Recent advancements focus on mitigating these issues through innovative material design and electrolyte optimization, aiming to improve the longevity and performance of Li–S batteries [10].

**Sodium–sulfur batteries (Na–S)** are promising candidates for next-generation grid-level storage systems due to their high theoretical capacity, low cost, and capability to provide stable and long-duration discharge. Traditionally, Na–S batteries operate at high temperatures (300–350°C), where sodium and sulfur become molten and facilitate ion transfer [11]. In contrast, room-temperature Na–S batteries have emerged as a viable alternative, taking advantage of sodium's natural abundance. However, the practical application of Na–S batteries is hindered by issues such as the polysulfide shuttle effect and sluggish reaction kinetics, leading to rapid capacity loss and performance degradation. Current research focuses on optimizing operating conditions and developing stable electrolytes to mitigate these challenges and enhance the performance of Na–S batteries [12].

**Metal–air batteries (MABs)** consist of four main components: a porous air cathode, a metal anode, an electrolyte, and a separator. By utilizing redox reactions between a metal (such as zinc or lithium) and oxygen from the air, MABs function as a hybrid of a battery and a fuel cell. They are considered a promising energy storage solution due to their low cost, high specific energy, high power density, and safety. However, the development of MABs faces significant challenges, including poor rate capability, dendrite formation, corrosion during electrochemical reactions, sluggish oxygen reaction kinetics at the cathode, and underdeveloped material design [13, 14].

**Redox-flow batteries (RFBs)** provide flexible and scalable solutions for stationary energy storage systems. Due to their reliance on stationary tanks for liquid electrolyte storage, RFBs are not suitable for mobile applications. They utilize liquid electrolytes stored in external tanks to store and release energy, allowing for independent scaling of energy and power. Highly adaptable, RFBs are well suited for large-scale stationary applications and can use various redox pairs to meet specific needs and optimize material costs. Despite these advantages, RFBs encounter challenges such as low energy density and high capital costs. Recent research has focused on improving membrane design, enhancing the performance of organic electrolytes, and developing new redox-active materials to address these issues. Advances in molecular design and modifications

144   D. S. Reichert & K. P. Birke

have notably enhanced the stability and efficiency of organic electrolytes employed in RFBs [15].

### 5.1.3 Raw materials and their significance for alternative battery technologies

The raw materials used in battery technologies play a crucial role in their performance, availability, cost, safety profiles, environmental friendliness, scalability, and application possibilities. Different materials offer varying theoretical capacities, voltage levels, and energy densities, which affect their suitability for different applications. In the following sections, we examine the influence and effects of the choice of ionic charge carrier on the respective battery technology. Table 5.1 gives a comparative overview of the physical properties of $Li^+$, $Na^+$, $Mg^{2+}$, $Zn^{2+}$, and $Al^3$ as charge carriers for rechargeable batteries.

**Performance:** The theoretical capacity of a material, measured in mAh/g, indicates how much electric charge can be stored per unit mass. Commonly used elements such as lithium (Li), sodium (Na), magnesium (Mg), zinc (Zn), and aluminum (Al) are selected for their high theoretical capacities. For instance, lithium has a theoretical capacity of about 3860 mAh/g, sodium offers about 1165 mAh/g, and magnesium about 2206 mAh/g. The volumetric capacities are particularly significant for evaluating energy storage capabilities. Lithium has a volumetric capacity of about 2061 mAh/cm³, while sodium's is lower at about 1129 mAh/cm³. The differences in capacity can be attributed to the atomic mass and ionic radius of the

**Table 5.1.**   Comparison of the physical properties of $Li^+$, $Na^+$, $Mg^{2+}$, $Zn^{2+}$, and $Al^3$ as charge carriers for rechargeable batteries [16, 17].

|  | $Li^+$ | $Na^+$ | $Mg^{2+}$ | $Zn^{2+}$ | $Al^{3+}$ |
|---|---|---|---|---|---|
| Atomic mass (u) | 6.94 | 23.00 | 24.31 | 65.41 | 26.98 |
| Ionic radius (Å) | 0.76 | 1.02 | 0.72 | 0.74 | 0.54 |
| Stokes radius in water (Å) | 2.38 | 1.84 | 3.47 |  | 4.39 |
| Stokes radius in PC (Å) | 4.8 | 4.6 |  |  |  |
| Gravimetric capacity (mAh/g) | 3860 | 1165 | 2206 | 820 | 2980 |
| Volumetric capacity (mAh/cm³) | 2061 | 1129 | 3834 | 5855 | 8046 |
| $E^0$ (V vs. SHE) | −3.04 | −2.71 | −2.36 | −0.76 | −1.68 |
| Desolvation energy in PC (kJ mol⁻¹) | 215.8 | 158.2 | 569.4 |  |  |

elements. For example, lithium, with an atomic mass of 6.94 u and an ionic radius of 0.76 Å, allows for denser packing of ions within electrode materials. Sodium, having an atomic mass of 23.0 u and an ionic radius of 1.02 Å, results in less dense packing and consequently lower energy density [18].

The electrochemical potential of the ionic charge carrier affects the voltage of the battery. At 3.04 V, lithium has the highest electrochemical potential among the potential candidates as a charge-carrier ion, which leads to a high theoretical battery voltage. The energy density of a battery, measured in Wh/kg or Wh/L, indicates the energy stored per unit volume or mass and is directly proportional to the material capacity and voltage. The storable energy ($E$) in a battery can be expressed by the following equation (5.1):

$$E = C \cdot V, \tag{5.1}$$

where $C$ is the gravimetric or volumetric capacity (in mAh/g or mAh/L) and $V$ is the voltage (in V). Since SIBs have both lower capacity and lower voltage compared to LIBs, their overall energy density is correspondingly lower. Multivalent cations, such as magnesium and aluminum, are very promising because they can release several electrons per ion, potentially resulting in higher capacities. However, their less negative electrode potentials often lead to lower energy densities (Wh/kg) of the overall system compared to monovalent cations such as lithium and sodium.

Ion mobility and diffusion rates are also critical, as they influence the battery's charging and discharging speeds. This underscores the distinct roles of electrolytes and electrodes. Electrolytes are designed to conduct ions while preventing the flow of electrons, thereby avoiding uncontrolled reactions during battery operation. Conversely, electrodes must conduct both ions and electrons to enable the recombination process as ions and electrons enter the electrodes. This leads to complex interactions with the charge-carrier ions, which vary in ionic radius, atomic mass, and charge density. For example, lithium ions diffuse faster through electrode materials than larger ions such as sodium or magnesium, directly impacting the battery's power density [19]. Aluminum, with a relatively small ionic radius of 0.54 Å and three positive charges, has a high charge density. This increases the desolvation energy barrier of $Al^{3+}$ in the electrolyte solution, making it difficult for the solvated aluminum ions to efficiently shed their solvent molecules and interact with the electrode surface. Efficient desolvation is crucial for high ion mobility and low internal resistance within the battery. The challenges in desolvation and ion transport currently limit the large-scale application of aluminum in battery systems, despite its high theoretical volumetric capacity [20].

**Availability and cost:** Materials that are abundant and cost-effective, such as sodium, aluminum, and zinc, reduce manufacturing costs and improve the economic feasibility of batteries. They also lower dependence on geopolitical risks [21], which is crucial for the large-scale deployment of energy storage systems. In contrast, materials such as lithium and cobalt, which are concentrated in specific regions, can lead to supply shortages and price volatility.

The **environmental** impact of mining and processing raw materials is a crucial factor. Efficient recycling of battery materials is essential for reducing environmental harm and enhancing sustainability. Materials that are easy to recycle minimize environmental damage and promote the long-term viability of battery technologies. For example, incorporating recycled materials into LIBs can reduce overall production costs and environmental impacts by up to 54% compared to using virgin materials [22, 23].

The use of less reactive and more stable raw materials can significantly enhance **battery safety**. For example, sodium and magnesium are less susceptible to thermal runaway compared to lithium, which makes them safer alternatives. Beyond the choice of charge carrier, the safety of a battery is also influenced by the cell chemistry and thermal stability of other active materials, including the electrodes and electrolyte.

**Scalability and application possibilities:** The ability to produce and deploy battery technology on a large scale is essential for its economic and practical application. Technologies that are easily scalable can more effectively meet the demand for large-scale energy storage systems. For example, SIBs are considered a promising alternative due to their scalability and cost-effectiveness for large-scale applications. Scalability in production technology also involves the drop-in capability of new battery chemistries into existing manufacturing lines. Ideally, new battery types should be compatible with current production processes to minimize costs and streamline adoption.

Given the numerous alternative battery technologies available, the SIB stands out as a particularly promising option for further investigation.

SIBs are currently attracting significant attention in research, industrialization, and scale-up efforts due to their unique advantages, including availability, safety, environmental friendliness, and overall strong performance. The availability and cost of raw materials are critical factors in the selection of battery technologies. Sodium, one of the most abundant elements in the Earth's crust, can be cost-effectively extracted from seawater. The low material costs and ease of sourcing sodium ensure the economic production of SIBs, making them particularly attractive for large-scale energy storage applications [24]. While ZIBs, MIBs, and AIBs have the potential for high energy densities, these technologies are still in the early stages of development and require significant adjustments to their cell chemistry and production processes [6, 7, 9, 25, 26]. In contrast, SIBs are considered a "drop-in" technology for existing LIB manufacturing lines, which simplifies the transition and reduces the need for extensive retooling [24, 27].

SIBs are particularly interesting due to their versatility across a wide range of applications. As shown in Figure 5.2, they are suitable for diverse use cases, including renewable and residential energy storage, commercial and industrial energy storage, electric two- and three-wheelers, and

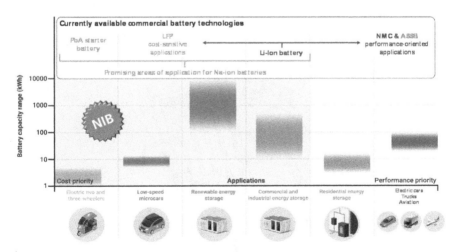

**Figure 5.2.** Currently available and near-future commercial battery technologies and their most promising application areas.

*Source*: Adapted from Ref. [32].

low-cost BEVs. This broad applicability makes SIBs a strong complementary or alternative technology to LIBs, setting them apart from other available technologies. In contrast, RFBs, MABs, and Na–S batteries, while promising in specific niches, generally have very specialized use cases and optimal performance conditions [12, 13, 28].

Due to their unique combination of properties, SIBs will be used in the following chapters to illustrate the trade-offs in selecting raw materials for battery technologies. This approach highlights the complexity of material selection, where various performance characteristics must be balanced. By focusing on SIBs, readers can gain a deeper understanding of the necessary compromises in developing new battery technologies.

## 5.2 Fundamentals: Lithium-Ion vs. Sodium-Ion Battery

### 5.2.1 *Basic principles of sodium-ion batteries*

SIBs operate on the same fundamental principle as LIBs, where ions move between the anode and the cathode to store and release electrical energy. This mechanism is often illustrated using the "rocking-chair model" which describes the movement of sodium ions (Na$^+$) between the anode and cathode during the charging and discharging cycles (see Figure 5.3). In this model, sodium ions shuttle back and forth, moving

**Figure 5.3.** "Rocking-chair model" of rechargeable batteries.

The Role of Raw Materials in Enhancing or Limiting Energy Density    149

from the cathode to the anode during charging and returning from the anode to the cathode during discharging. This continuous ion movement is essential for the storage and release of energy in the battery.

During the charging process, sodium ions are released from the cathode and travel through the electrolyte to the anode, where they are intercalated (inserted). For example, a common cathode reaction is

$$Na_xMO_2 \rightarrow Na_{x-y}MO_2 + yNa^+ + ye^-, \tag{5.2}$$

while an anode reaction might be

$$yNa^+ + ye^- + C \rightarrow Na_yC. \tag{5.3}$$

During the discharging process, the process reverses. Sodium ions are released from the anode, travel back through the electrolyte to the cathode, and release electrons that flow through the external circuit to perform work. A typical anode reaction during discharge is

$$Na_yC \rightarrow yNa^+ + ye^- + C, \tag{5.4}$$

while the cathode reaction is

$$yNa^+ + ye^- + Na_{x-y}MO_2 \rightarrow NaxMO_2. \tag{5.5}$$

The most common method of charge storage in SIBs is intercalation, where sodium ions are inserted into the layered structures of anode and cathode materials. Alternatively, conversion and alloying materials can be used, involving chemical reactions between sodium ions and electrode materials. While these mechanisms often offer higher capacities, they are associated with larger volume changes and mechanical stress, which can affect cycle life [29].

Ongoing research and development in this field aims to increase the capacity and voltage potential of sodium-ion electrode materials to enhance their energy density and overall competitiveness with LIBs. This research encompasses new materials and electrolyte formulations, as well as the overall interaction of all active components [30, 31]. The formulation of the electrolyte is particularly crucial for the formation of the solid–electrolyte interphase (SEI) in SIBs, similar to LIBs. The SEI layer forms during the initial charging cycles and performs several important functions. First, it acts as a passivation layer, preventing further electrolyte decomposition by blocking direct contact between the electrolyte and the electrode material. This protection is essential for maintaining the stability and longevity of the battery. Second, the SEI layer facilitates ion transport while blocking electron flow, which helps reduce self-discharge and improves overall efficiency. Third, a stable SEI layer can help to mitigate volume changes in the electrode materials, reducing mechanical stress and enhancing cycle life. Therefore, optimizing electrolyte formulations to promote the formation of a stable and robust SEI layer is a key area of research aimed at improving the performance and energy density of SIBs [32, 33].

### 5.2.2 *Energy density*

Energy density is a crucial property of batteries, as it determines the amount of energy stored per unit volume or mass. This section examines the differences in energy density between LIBs and SIBs, focusing on factors such as atomic mass, ionic radius, and other relevant aspects.

The **atomic mass** of the ions significantly influences the specific capacity of batteries. Lithium, with an atomic mass of 6.94 u, has a high theoretical specific capacity of approximately 3860 mAh/g. This high capacity contributes to the elevated potential energy density of LIBs, allowing them to store more energy per unit mass. In contrast, sodium has a higher atomic mass of 22.99 u, leading to a lower specific capacity of about 1165 mAh/g. This results in a reduced gravimetric energy density for SIBs compared to LIBs.

The **ionic radius** also significantly affects the energy density of batteries. Lithium ions ($Li^+$), with a small ionic radius of about 0.76 Å, can efficiently intercalate into the crystalline structures of electrode materials. This small size allows LIBs to achieve high energy densities, as the smaller ions can be packed more densely within the electrode materials, enhancing the volumetric energy density. In contrast, sodium ions ($Na^+$) have a larger ionic radius of approximately 1.02 Å. This larger ionic radius necessitates greater interlayer distances in layered materials or larger defect sites for accommodation, which reduces the volumetric density of the active materials and consequently lowers the overall energy density of SIBs. Specifically, the theoretical volumetric capacity of lithium is approximately 2061 mAh/cm³, whereas that of sodium is about 1129 mAh/cm³.

The **electrochemical potential** of ions impacts cell voltage and, consequently, the energy density of batteries. The electrochemical potential

$E^0$ (V vs. SHE) of the charge-carrying ion indicates the energy change associated with the ion's reduction or oxidation. Higher electrochemical potentials contribute to higher cell voltages, which can enhance energy densities. For example, lithium ions ($Li/Li^+$) possess an electrochemical potential of $-3.04$ V, enabling LIBs to achieve high voltages. This high potential contributes to the high energy densities observed in LIBs, particularly with cathodes such as nickel manganese cobalt oxides (NMC 622 and NMC 811), which typically operate within a voltage range of 2.5–4.2 V [34, 35]. In contrast, sodium ions ($Na/Na^+$) have a lower electrochemical potential of $-2.71$ V. This results in a lower potential cell voltage for SIBs. For example, SIBs with nickel–manganese–iron oxide (NFM) cathodes generally operate within a voltage range of 2.0–4.0 V, leading to a reduced overall energy density compared to LIBs [36].

Lithium's higher electrochemical potential and therefore greater reduction potential result in a higher reactivity compared to sodium. This increased reactivity facilitates the formation of a stable SEI layer on the anode surface during the initial charging cycles. The SEI layer acts as a protective passivation layer, preventing continuous electrolyte decomposition and ensuring the stability of the anode material. This stability is crucial for maintaining the high energy density observed in LIBs [37]. In contrast, the SEI layer in SIBs encounters significant stability challenges compared to LIBs. Due to the higher solubility of SEI components, such as sodium fluoride (NaF) and sodium carbonate ($Na_2CO_3$), in typical electrolyte solvents, the SEI layer is often more dynamic and less robust, leading to higher internal resistance and reduced efficiency [38].

Viewed in isolation, specific active materials or electrolytes may appear to promise high theoretical energy densities per mass or volume. However, these figures often overlook many practical considerations and cannot be directly transferred to real values. The disparity between theoretical and practical capacities and voltages can be significant across different battery systems. Understanding the underlying electrochemical processes is crucial for accurately estimating the usable energy content of a battery. A pertinent example is lead–acid batteries (LABs). While the specific energy based on solid active materials (Pb anode and $PbO_2$ cathode) is around 240 Wh/kg, this value is reduced to 166 Wh/kg when the required sulfuric acid electrolyte is included. In practical applications, such as in motor vehicles, LABs achieve an energy density of around 35 Wh/kg. This reduction is due to factors such as electrolyte dilution, low utilization of the active mass, and the weight of inactive components.

This example illustrates a common issue in battery research: Theoretical energy values can be misleading if not all contributing materials are considered. Comprehensive and transparent energy calculations are essential to accurately assess and compare the potential of different battery technologies [39].

### 5.2.3 *Power density*

Power density refers to the amount of power that a battery can deliver per unit volume or mass and is crucial for determining how quickly a battery can provide energy. It is influenced by several factors, including the rates of electron and ion diffusion within the battery materials as well as the complex interplay of ionic properties, solvation dynamics, and material interactions. These factors determine the speed at which a battery can charge and discharge, affecting its overall performance and suitability for applications with high power demands, such as electric vehicles and portable electronics. The choice of the primary ion, whether lithium (Li) or sodium (Na), significantly influences power density through various mechanisms and material properties.

Ion mobility and diffusion rates are critical as they influence the charging and discharging speeds of batteries. This highlights the distinct roles of electrolytes and electrodes. Electrolytes are designed to conduct ions while inhibiting the flow of electrons, thereby preventing uncontrolled reactions during battery operation. Conversely, electrodes must facilitate the conduction of both ions and electrons to enable the recombination process as ions and electrons enter the electrodes.

The smaller **ionic radius** (Shannon's ionic radius) and lower **atomic mass** of lithium ions facilitate faster ion diffusion within electrode materials. This enhanced movement of lithium ions through the electrode potentially results in quicker charge and discharge rates, leading to higher power densities. In contrast, SIBs face challenges due to the larger ionic radius and higher atomic mass of sodium ions, which result in slower ion diffusion within the electrode [40].

In the electrolyte, however, the situation is different. The desolvation process at the electrode–electrolyte interface, where charge-carrier ions shed their solvent molecules to efficiently interact with the electrode surface, is known to be a rate-determining step. Figure 5.4 illustrates the Stokes radii (hydrodynamic radius of the solvated ion) of various ions in propylene carbonate (PC). The trend shows that the higher the surface

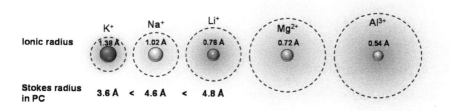

**Figure 5.4.** Comparison of Shannon's ionic radii and Stokes radii (in PC) among $Li^+$, $Na^+$, $K^+$, $Mg^{2+}$, and $Al^{3+}$ ions.

*Source:* Adapted from Ref. [39].

charge density of the ion and the stronger its Lewis acidity, the stronger the interactions with polar solvent molecules. This leads to a larger Stokes radius for lithium ions compared to sodium ions. Consequently, sodium ions experience lower desolvation energy in PC compared to lithium ions, making them more easily desolvated and intercalated into the electrode material. This lower desolvation energy contributes to the improved ionic conductivity of sodium ions in certain electrolytes, potentially enhancing the power density of SIBs [41, 42].

**Crystal structure and volume changes:** The smaller ionic radius of lithium ions theoretically leads to less volume change in electrode materials during charge and discharge cycles, contributing to greater mechanical stability and longer battery life. This stable intercalation process enhances the power density of LIBs. In contrast, the larger sodium ions can cause significant volume changes, which may cause mechanical stress and potential cracking in electrode materials. Such mechanical degradation reduces the cycle life and power density of batteries, as the structural instability impairs the electrodes' ability to sustain rapid ion movement [43, 44].

**SEI formation:** The SEI formed during the initial cycles serves as a passivation layer, regulating the flow of ions and consequently increasing the battery's impedance and resistance. This layer is vital for the electrode's chemical and mechanical stability. It must effectively passivate the electrode surface while allowing lithium ions to pass through, thereby significantly reducing undesirable electron transfer. As discussed in Section 5.2.2 Energy density, the higher solubility of the SEI in SIBs can

## 5.2.4 *Stability and safety*

The stability and safety of batteries are critical factors influenced by the specific properties of the ions used. This section explores the key differences between lithium-ion ($Li^+$) and sodium-ion ($Na^+$) batteries regarding their effects on degradation mechanisms, lifespan, and overall safety.

**Degradation mechanisms:** Battery degradation refers to the gradual decline in a battery's performance and capacity over time due to various physical and chemical changes within its components. These changes can include alterations in the electrode materials, electrolyte decomposition, formation of passivation layers, mechanical stresses, and other phenomena that collectively diminish the battery's ability to store and deliver energy effectively. These degradation mechanisms can be divided into primary and secondary effects [46].

Primary degradation mechanisms include SEI layer growth, particle fracture, metal plating, and structural disorder.

**SEI layer growth:** The SEI layer forms on the surface of the negative electrode, providing a passivation layer that prevents further electrolyte decomposition. However, the continuous growth of the SEI layer consumes cyclable lithium or sodium ions, leading to capacity loss and increased resistance. This SEI layer initially forms during the first cycle, causing an initial capacity loss but then stabilizes to protect the electrode. However, over time, factors such as high temperatures and cycling can lead to further SEI growth, increasing internal resistance and reducing the overall performance of the battery [46].

**Particle fracture:** Electrode particles can fracture due to significant volume changes during cycling, especially in high-capacity materials such as silicon. This mechanical failure exposes new surfaces to the electrolyte, exacerbating SEI growth and further reducing capacity. Additionally, the fracture of electrode particles can disrupt electrical connectivity within the electrode, leading to isolated areas of active material that no longer contribute to the battery's capacity [46].

**Lithium plating:** At high current densities and low temperatures, metallic lithium can deposit on the anode surface instead of intercalating into the electrode. This can lead to the formation of dendrites, which can pierce the separator and cause internal short circuits, severely compromising battery safety. Lithium plating occurs more readily at low temperatures or high charging rates, where the anode becomes fully lithiated or sodiated and unable to accommodate more ions [46].

**Positive electrode structural change and decomposition:** The crystal structure of positive electrode materials can degrade over time. Transition metal oxides, used in cathode materials, can undergo phase transitions, oxygen release, and form less conductive phases such as spinel and rock salt. These changes impede ion intercalation and deintercalation, reducing the amount of active material available and increasing the internal resistance of the battery [46].

Secondary degradation mechanisms are a consequence of the primary ones and include graphite exfoliation, pore blockage, island formation, dissolution of transition metals, electrolyte decomposition, gas formation, HF formation, and the formation of a positive SEI layer (pSEI).

**Graphite exfoliation:** Mechanical degradation can cause the graphite layers in the negative electrode to exfoliate, reducing the effective surface area for lithium-ion intercalation and leading to increased resistance and capacity loss. This process can be exacerbated by repeated cycling and high currents, which increase mechanical stresses within the electrode [46].

**Pore blockage:** The growth of the SEI layer and other side reactions can block the pores in the electrodes, reducing the effective surface area for charge-carrier ion transport and increasing the cell's internal resistance. Pore blockage restricts the flow of electrolyte through the electrode, impeding ion transport and reducing the overall performance of the battery [46].

**Island formation:** Fractured electrode particles can form isolated islands that are no longer electrically connected, leading to a loss of active material and decreased battery capacity. These isolated islands result from mechanical stresses and volume changes during cycling, which break up the electrode structure [46].

**Transition metal (TM) dissolution:** Transition metals from the positive electrode can dissolve into the electrolyte and migrate to the negative electrode. This migration can lead to additional SEI growth and increased impedance, further reducing battery performance. TM dissolution is often accelerated by high temperatures and high voltages, which increase the reactivity of the cathode material [46].

**Electrolyte decomposition:** High voltages, high temperatures, and side reactions can cause the electrolyte to decompose, leading to gas formation, electrolyte loss, and increased cell impedance. Electrolyte decomposition can produce gases such as $CO_2$, which increases the internal pressure of the cell and can lead to swelling and leakage [46].

**Gas formation:** Electrolyte decomposition and other side reactions can produce gases that increase the internal pressure of the cell, potentially leading to swelling, leakage, and reduced cell performance. Gas formation is often associated with high operating temperatures and overcharging, which increase the rate of side reactions [46].

**HF formation:** Electrolyte decomposition of salts such as $LiPF_6$ can produce hydrofluoric acid (HF), which can further degrade the electrodes and the SEI layer, leading to accelerated capacity loss and increased impedance. HF is highly reactive and can cause significant damage to both the electrode materials and the SEI layer [46].

**pSEI layer formation:** A passivation layer similar to the SEI on the negative electrode can form on the positive electrode, known as the positive SEI (pSEI) layer. This pSEI layer can impede lithium-ion transport, contributing to increased impedance and capacity fade. The pSEI layer forms as a result of transition metal dissolution and side reactions with the electrolyte, and its growth can reduce the overall performance of the battery [46].

Although sodium-ion and lithium-ion systems have similar components and operating principles, they exhibit some distinct differences in aging and degradation mechanisms. For example, sodium-ion cathodes generally exhibit greater thermal stability compared to lithium-ion cathodes. However, the SEI in sodium-ion systems tends to degrade more rapidly [47].

**Electrode material degradation:** In SIBs, the larger sodium ions can induce more significant volume changes in electrode materials during charge and discharge cycles. Both anodes and cathodes in SIBs experience mechanical stress due to expansion and contraction during cycling. Hard carbon (HC) anodes expand less than graphite but still encounter structural challenges. Conversion and alloying materials undergo significant expansion, leading to particle cracking and poor cycle life. Cathode materials, particularly layered oxides, suffer from lattice strain, structural collapse, and amorphization, which contribute to performance degradation [47].

**SEI and CEI instabilities:** The SEI on HC anodes and the cathode–electrolyte interphase (CEI) are crucial for cell stability. The SEI in SIBs is less stable, more soluble, rougher, and less uniform compared to that in LIBs and is composed of species such as NaF and $Na_2CO_3$. A less stable SEI leads to continuous decomposition, increased resistance, and reduced cycle life. Similarly, the CEI suffers from instability, which further degrades the cathode materials [47].

**Sodium plating and dendrite formation:** Sodium plating is more likely to occur on HC anodes, particularly under high current densities or overcharge conditions. This phenomenon leads to the formation of dendrites – needle-like structures of sodium metal – that can penetrate the separator, creating internal short circuits. These short circuits can cause localized heating, SEI creation, electrolyte decomposition, and increased interfacial resistances, significantly impacting the safety and longevity of SIBs. This poses a severe risk of thermal runaway, fire, and explosion [47].

**Thermal stability:** Sodium-ion cathodes generally exhibit greater thermal stability than their lithium counterparts, reducing the likelihood of rapid thermal runaway. This higher thermal stability is attributed to the structural resilience of the cathode materials, which can withstand higher temperatures without significant degradation. However, despite this advantage, the SEI in SIBs is less stable and begins to decompose at lower temperatures compared to that in LIBs [47].

**Cell design and material choices:** The use of aluminum for both the anode and cathode current collectors in SIBs significantly enhances safety compared to LIBs, which use copper for the anode current collector. In LIBs, copper can dissolve at low cell voltages, leading to the formation of copper dendrites that pose a risk of short circuits. Aluminum does not dissolve under the operating conditions of SIBs, eliminating the risk of dendrite formation from current collector dissolution. SIBs can be safely discharged to and stored and transported at 0 V, reducing the risk of short circuits and thermal runaway during handling. This feature enhances the overall safety of SIBs compared to LIBs [47].

## 5.3 Sodium-Ion Battery: Design Principles for Electrode Materials

The development of high-performance and long-lasting SIBs heavily relies on the choice and optimization of electrode materials. These materials significantly influence the overall performance, stability, and lifespan of the batteries. Optimizing electrode materials for SIBs involves not only intrinsic properties but also interactions with the electrolyte and structural integrity over numerous charge–discharge cycles. In SIBs, as with LIBs, the electrodes play a crucial role in facilitating the reversible intercalation and deintercalation of ions during charging and discharging cycles. These processes involve redox reactions, where the oxidation of the anode material releases electrons and sodium ions, while the reduction of the cathode material absorbs these electrons and incorporates the ions. The electrode materials serve as hosts for the ions and provide pathways for electron transfer, which are essential for storing and releasing energy.

This section delves into the fundamental design principles for electrode materials in SIBs, focusing on factors such as size, morphology, structural effects, and transport properties. These principles are crucial for maximizing electrochemical performance and addressing challenges associated with sodium-ion technology.

### 5.3.1 Transport properties

The transport properties of SIBs are essential for enhancing energy density, power density, and cycling life. These properties involve the movement of sodium ions and electrons both within the bulk electrodes and at the interfaces between electrodes and electrolytes. Key parameters that reflect these behaviors include the chemical diffusion coefficient, electronic conductivity, and ionic conductivity. Effective control and optimization of these transport properties are essential for improving the electrochemical performance of SIBs. The chemical diffusion coefficient is a critical parameter that determines the movement of electrons and ions within a concentration gradient. In SIBs, the diffusion of sodium ions within the electrode material often represents the rate-determining step in various electrochemical processes. A higher diffusion coefficient indicates faster ion transport, which leads to enhanced rate capabilities and improved power density [48, 49].

**Doping** is a widely used approach to enhance the transport properties of SIBs by modifying the chemical diffusion coefficient, electronic conductivity, and ionic conductivity, leading to significant improvements in electrochemical performance. Research has demonstrated that doping $Na_3V_2(PO_4)_3$ with various elements, such as Li, K, Mg, Ni, Fe, Cu, Al, Mo, Cr, Ti, and Mn, can optimize the crystalline structures, transport properties, and electrochemical performance of the electrodes [48]. For instance, both cation and anion doping have significantly impacted the structural properties of polyanionic electrodes for SIBs, resulting in extraordinary electrochemical performance [50]. Moreover, dual doping strategies, such as incorporating Ti and V into $Na_3Ti_{0.5}V_{0.5}(PO_3)_3$, have been used to elevate the redox potential of titanium-based cathodes, leading to improved cycling stability and distinctive voltage platforms [51]. Elemental doping has also proven effective in mitigating irreversible phase transitions and improving the cycling performance of manganese-based layered oxide materials in SIBs [52]. Additionally, Sc and Ge co-doping in $Na_3V_2(PO_4)_3$ enhances both bulk and grain boundary conductivity, yielding impressive cycling capacities [53]. Further, Zn-doped $Na_4MnV(PO_4)_3$, combined with a carbonized polyacrylonitrile coating layer, has shown improved discharge capacity and cycling stability [54]. Moreover, Nb doping in $Na_3V_2(PO_4)_3$ enhances air stability and improves electrochemical performance by forming a stable CEI film [55]. Overall, doping strategies involving elements such as phosphorus and zinc have demonstrated significant enhancements in the electrochemical performance of SIBs, improving ion transport, stability, and capacity retention [56–58].

**Carbon coating:** Carbon coating is a highly efficient strategy for enhancing the transport properties and electrochemical performance of electrode materials with low electronic conductivity. This method improves electronic conductivity by providing a conductive network that facilitates better electron transport. For instance, the electrochemical performance of $Na_4MnV(PO_4)_3$ was significantly improved through a double-carbon-layer coating strategy. This approach resulted in enhanced electronic conductivity, structural stability, and excellent sodium storage capabilities [59]. Similarly, a multifunctional coating applied to a sulfur-containing carbon-based anode improved sodium storage performance by enhancing capacity and cycling stability due to the combined physical and chemical effects of the coating [60]. In another instance, the synergistic effect of a 3D electrode architecture and an *in situ* carbon coating on $SnO_2$ anodes significantly advanced the electrochemical performance of SIBs, demonstrating stable capacity retention and improved C-rate performance [61].

### 5.3.2 *Active redox sites*

The materials selected for electrodes must effectively support redox reactions during both charging and discharging processes. Efficient electrode materials should enable a high packing density of ions while stabilizing them within the electrode structure. This stabilization is crucial for preserving the structural integrity of the electrode during cycling, thereby ensuring long-term durability and consistent energy density. Additionally, the efficiency of redox reactions is vital for achieving high energy density.

Transition metals are commonly employed in cathode materials because of their ability to easily switch oxidation states, which facilitates redox reactions. In LIBs, transition metals such as iron, cobalt, nickel, and manganese are frequently used. These metals form compounds like $LiFePO_4$ (LFP) and $LiNiMnCoO_2$ (NMC), which are favored for their high specific capacities and their ability to maintain structural integrity during numerous charge–discharge cycles [62].

In contrast, SIBs typically use transition metals such as iron, manganese, and vanadium, forming compounds like $Na_3V_2(PO_4)_3$ and $Na_{0.7}MnO_2$. While these materials share structural similarities with their lithium counterparts, they generally offer lower specific capacities and energy densities. This decrease is primarily due to the larger size of sodium ions, which affects the structural stability of the electrode materials [63].

The choice of transition metals significantly impacts the energy density of the battery. Cobalt-based materials, for instance, offer high energy density but are expensive and face supply chain challenges. Nickel and manganese provide a good balance between energy density and cost, although they require careful management to avoid degradation. In SIBs, iron-based materials are cost-effective and environmentally friendly but typically have lower energy densities. Manganese and vanadium can enhance performance but also present challenges in terms of material stability and energy density [64].

### 5.3.3 Size effects

The size of the active material particles in the electrodes significantly impacts the performance of SIBs. Smaller particles provide a larger surface area, which enhances the number of sites available for electrochemical reactions, leading to improved charge and discharge rates. For spherical particles in the active material, the ratio of surface area to volume increases as the particle size is reduced. With a constant mass of active material, the surface area increases as the particle size decreases.

**Surface area:** A larger specific surface area enhances reactivity, which can improve both capacity and cycle life. An increased surface area facilitates better interaction between the electrode materials and the electrolyte, promoting efficient ion diffusion and electrochemical reactions. However, an excessively large surface area can also lead to undesirable side reactions with the electrolyte, potentially compromising stability. Therefore, it is crucial to balance surface area with stability to achieve optimal performance [65, 66].

**Nanostructuring:** Reducing particle size to the nanometer range shortens the diffusion path for sodium ions and electrons, increasing charge and discharge rates. Nanostructured materials often exhibit better electrochemical performance due to their larger specific surface area and improved kinetics. For example, nanostructuring $NaTi_2(PO_4)_3@C$ particles embedded into nitrogen-doped graphene sheets has demonstrated enhanced electronic conductivity and improved electrochemical performance, including high reversible capacity and stable cyclic properties [67]. However, extremely small particles can lead to increased electrolyte decomposition. Despite this, the higher specific surface area of nanostructured materials generally results in greater reactivity, which can improve capacity and cycle life – key factors for the efficiency and performance of SIBs [66, 68].

**Particle agglomeration:** Nanostructured materials are prone to agglomeration, which can reduce the benefits of nanostructuring. Agglomeration reduces the effective surface area and can impair battery performance.

To mitigate this issue, techniques such as using appropriate binders and additives are essential for maintaining particle stability and distribution, thereby maximizing the advantages of nanostructuring [69].

**Particle size distribution:** A uniform particle size distribution is critical for ensuring consistent reactions throughout the electrode. Variations in particle size can lead to inhomogeneous reaction rates and inconsistent aging of the battery. Advances in manufacturing techniques, such as controlled synthesis methods and technologies like spray pyrolysis or chemical vapor deposition, can improve particle size distribution and optimize battery performance [70].

**Influence of particle size on electrode stability:** Smaller particles generally offer greater mechanical stability and are less likely to crack or experience structural changes during charge–discharge cycles. This improved stability can extend the battery's lifespan and maintain its performance over many cycles [71].

**Optimization of electrolyte compatibility:** The compatibility of electrode materials with the electrolyte is another critical factor in battery performance. Nanostructured materials can increase the surface area available for interaction with the electrolyte, leading to better ion diffusion and a more stable electrolyte–electrode interface [72].

### 5.3.4 *Structure and morphology*

The morphology and structure of electrode materials play a central role in the performance and longevity of SIBs. These properties influence ion diffusion, mechanical stability, and the electrochemical reactivity of the materials.

**Structure of electrode materials:** The crystalline structure of electrode materials determines their ion diffusion properties and stability during charge–discharge cycles. Materials with well-ordered crystalline structures offer clear diffusion paths for sodium ions, enhancing charge and discharge rates. For example, layered and tunnel structures are particularly advantageous for sodium-ion diffusion [73]. In contrast, amorphous and semicrystalline structures can increase capacity by providing more intercalation sites for sodium ions, although they tend to be less stable and may degrade more quickly with repeated cycling [74].

Studies have shown that the crystalline structure of bicrystalline titanium dioxide spheres with a core–shell structure significantly impacts ion diffusion in SIBs. These structures demonstrate high performance in terms of initial discharge capacity, cycle stability, and rate capability [75]. Additionally, materials with intercalation-type reaction mechanisms, such as crystalline iceplant-like nano-NaVPO$_4$F@graphene structures, exhibit excellent cycling and rate performance due to their crystalline nature [76].

Moreover, high crystalline Prussian white nanocubes have shown promising performance as cathode materials for SIBs, emphasizing the importance of crystalline structure in enhancing ion diffusion and battery performance [77].

**Influence of morphology on electrochemical performance:** Recent advancements have focused on improving power density in SIBs by developing novel cathode materials with multidimensional Na$^+$ migration pathways. Notable examples include Prussian blue analogs and NASICON-type materials with three-dimensional Na$^+$ diffusion pathways, which are crucial for constructing high-power-density SIBs [40, 78].

A larger specific surface area improves the interaction between electrode materials and electrolytes, enhancing capacity and cycle life. However, an excessively large surface area can lead to undesirable side reactions. Nanostructured materials are also at risk of agglomeration, which can diminish the benefits of their nanostructure. Techniques to prevent agglomeration, such as using appropriate binders and additives, are essential. For instance, sodium alginate (SA) as a binder significantly improves electrode stability and kinetics [79].

**Porosity:** High porosity in electrode materials can enhance ion diffusion and performance. Porous structures allow better electrolyte penetration and facilitate ion transport, thereby improving charge and discharge rates. However, excessive porosity can lead to structural weaknesses and reduced mechanical stability. It is crucial to balance adequate porosity with structural strength. For instance, the varying porosity in the jute fiber precursor-derived HC anode impacts the reversible capacity and capacity retention of SIBs, with mesoporous HC outperforming ultramicroporous HC [80].

**Binders and additives:** The choice of binders and additives can affect the mechanical stability and conductivity of electrode materials. Suitable binders such as carboxymethylcellulose (CMC) and additives like carbon black can improve electrode performance and stability by keeping particles well bound and enhancing conductivity. Additionally, binders and additives play a key role in preventing particle agglomeration, supporting the effectiveness of nanostructuring. Studies have shown that water-soluble inorganic binders containing Li and Na phosphates and silicates are particularly effective due to their intrinsic ionic conductivity and compatibility with a wide range of electrode materials [81].

**Mechanical properties:** The mechanical properties of electrode materials, including elasticity and strength, play a critical role in resisting volume changes during charge–discharge processes. Materials that are both flexible and strong can better absorb mechanical stress, extending battery life. For example, the use of cross-linked SA/graphene oxide (GO) binders has demonstrated improved rate capability and cycling stability compared to traditional binders such as PVDF and pure SA [82].

### 5.3.5 *Defect engineering*

Defect engineering is a relevant strategy for optimizing the performance of electrode materials in rechargeable batteries, including SIBs. Defects in the crystal structure of electrode materials can significantly influence their electrochemical properties by providing additional active sites, enhancing ion diffusion, and improving electronic conductivity. Introducing defects into these materials can create more active sites for ion storage, enhance ion and electron transport, and stabilize the structure during cycling. This stabilization is essential to maintain the electrode's integrity, thereby improving the overall efficiency and lifespan of the battery.

**Types of defects and their effects:** Intrinsic defects, such as Schottky and Frenkel defects, naturally occur due to thermal vibrations of lattice atoms. Schottky defects involve vacancies in the lattice, while Frenkel defects result from atoms or ions moving into interstitial sites, creating interstitial atoms. Non-intrinsic defects, induced by impurity atoms or ions embedded in the lattice, are known as impurity defects. These defects, such as heteroatom doping, can modify the electronic structure and chemical properties of the material, enhancing its electrochemical performance. For example, oxygen vacancies in transition metal oxides can enhance ion diffusion and charge transfer by creating low-energy pathways for ion transport [83].

**Defect engineering strategies:**

1.  *Chemical reduction and etching*: This method involves high-temperature treatments in reducing atmospheres to introduce oxygen vacancies in materials. For example, titanium dioxide ($TiO_2$) can be treated in a reducing atmosphere to create oxygen vacancies, which improve its electrochemical properties by enhancing ion diffusion and electronic conductivity [84]. The chemical reduction process can also involve the use of reducing agents, such as sodium borohydride or hydrazine hydrate, to induce vacancy defects. These vacancies act as active sites for ion storage and facilitate faster ion transport, thereby enhancing the overall performance of the electrode material.

2.  *Physical exfoliation and etching*: Techniques such as plasma technology and mechanical ball milling are used to create defects in nanomaterials. Plasma technology, for instance, can introduce defects in carbon-based materials without damaging their overall structure, thereby enhancing their electrochemical performance [79]. Mechanical ball milling is effective for generating vacancies and dislocations in electrode materials, increasing their surface area and reactivity. This method is particularly effective for producing nanostructured materials with high specific surface areas, which are beneficial for ion storage and transport.

3.  *Electrochemical treatment*: Applying an electric field to electrode materials in an electrochemical cell can induce defects by driving ions into or out of the material. This method can create vacancies and interstitials that enhance ion diffusion and storage capacity.

4.  *Annealing and quenching*: Rapid thermal treatments, such as annealing followed by quenching, can introduce a high concentration of defects in electrode materials. These treatments can induce phase transformations and create metastable phases, resulting in enhanced electrochemical properties.

**Impact on performance:** *Ion storage and diffusion*: Defects such as heteroatom doping and vacancies in nanostructured materials can enhance ion diffusion and electron transfer, leading to improved conductivity and rate performance of the electrode materials. For example, oxygen

166  *D. S. Reichert & K. P. Birke*

vacancies in titanium dioxide increase its conductivity and accelerate ion dynamics, enhancing its performance in SIBs [85].

Defects can also improve the structural stability of electrode materials during cycling. For example, vacancy-containing structures can maintain their integrity during ion insertion and extraction, reducing the likelihood of structural collapse and enhancing the battery's lifespan. This effect has been demonstrated in materials such as $Na_{4/7}[Mn_{6/7}(VMn)_{1/7}]O_2$, which exhibits enhanced structural stability due to the presence of periodic vacancies [86].

## 5.4 Cathode Materials

The choice of cathode materials is crucial for the performance of SIBs. These materials must possess several key characteristics, including high capacity, good cyclability, thermal stability, and safety. Additionally, they should enable efficient sodium-ion intercalation and deintercalation, maintain structural integrity during cycling, and provide a high operating voltage to ensure a high energy density. Various material classes have been explored to meet these requirements, each offering a unique set of advantages and challenges.

### 5.4.1 *Transition metal oxide cathodes*

Transition metal oxides are among the most promising cathode materials for SIBs due to their relatively high theoretical capacity and good cycle life. Sodium transition metal oxides ($Na_xMO_2$) are particularly notable for their efficient intercalation and deintercalation of sodium ions. Examples include $Na_{0.7}CoO_2$, $Na_{0.7}MnO_2$, and $Na_{0.7}FeO_2$, which provide high capacity and good cycle stability [87].

The structure and diffusion properties of sodium transition metal oxides are crucial for their performance. These materials have a layered structure that promotes good sodium-ion diffusion, enhancing the mobility of sodium ions and ensuring efficient charge and discharge performance. This layered structure ensures high specific capacities, although stability can be challenged by phase transitions and structural changes during cycling. For instance, $NaFeO_2$ is valued for its stable layered structure and high capacity, whereas $NaMnO_2$ is appreciated for its high capacity and cost-efficiency. However, $NaNiO_2$ and $NaCoO_2$, while offering high energy densities, face limitations due to high costs and material toxicity [87–89].

**Phase transitions in transition metal oxides:** Phase transitions involve changes in the crystal structure of the material during sodium-ion intercalation and deintercalation. These transitions are critical because they can significantly impact the material's performance and longevity. The most common phases observed in SIB cathode materials are the O3, P2, P3, and O2 phases:

**O3 phase:** In this phase, sodium ions are located in octahedral sites between the layers of transition metal oxides. The O3 phase is known for its high capacity and good stability during the initial stages of cycling. However, repeated cycling can cause phase transitions to other structures, leading to volume changes and mechanical stress. These changes can reduce the material's structural integrity and overall battery performance [90].

**P2 phase:** This phase features prismatic coordination sites for sodium ions. It generally offers better stability during cycling compared to the O3 phase, providing improved long-term performance and stability. The P2 structure facilitates smoother sodium ion diffusion, reducing the likelihood of significant volume changes and mechanical degradation. Materials with a P2 structure, such as $Na_{0.7}MnO_2$, often exhibit enhanced cycle life and stability [91].

**P3 phase:** The P3 phase is similar to the P2 phase but has a different arrangement of sodium ions and transition metal layers. It aims to balance the high capacity of the O3 phase with the stability of the P2 phase. Despite its benefits, the P3 phase can still undergo phase transitions during cycling, potentially affecting performance. Although less common, the P3 phase offers unique advantages for specific applications [92].

**O2 phase:** In the O2 phase, sodium ions occupy octahedral sites, similar to the O3 phase, but with a different stacking sequence of the oxide layers. This phase provides a compromise between the structural stability of the P2 phase and the high capacity of the O3 phase. The O2 phase can offer improved ionic conductivity and structural integrity during cycling. However, like other phases, it remains susceptible to phase transitions, which can lead to volume changes and mechanical stress [93].

There exist several strategies to address these phase transition issues. **Mixed metal oxides:** Combining multiple metals in the oxide structure can enhance the stability of the crystal structure and mitigate the effects of phase transitions. For example, materials such as $NaNi_xMn_yCo_zO_2$ (NMC) and $NaFe_yCo_zO_2$ (NFC) incorporate various metals to achieve a more stable structure, distributing mechanical stress more evenly and reducing the likelihood of catastrophic failure during phase transitions [91].

**Surface coatings:** Applying protective coatings to electrode particles serves as a buffer layer that absorbs mechanical stress and prevents direct contact between the electrolyte and the active material. This technique reduces side reactions and improves overall stability and lifespan. Common coating materials include metal oxides, phosphates, and carbon-based materials [94].

**Nanostructuring:** Designing electrode materials at the nanoscale can improve their tolerance to phase transitions. Nanostructured materials can better accommodate volume changes compared to bulk materials due to their higher surface area-to-volume ratio and the presence of more grain boundaries, which serve as sites for stress relief. Additionally, the smaller particle size reduces the diffusion paths for sodium ions, thereby improving the overall kinetics of the battery [95].

### 5.4.2 *Polyanion-type cathodes*

Polyanion-type cathodes are highly valued for their high thermal stability and safety, making them particularly suitable for applications where these characteristics are critical. These materials are characterized by robust frameworks that contribute to their stability during electrochemical cycling and relatively high operating voltages compared to other cathode materials. Polyanion cathodes derive their stability and high voltage from the presence of polyanion groups (e.g., $PO_4$ and $SO_4$) within their structure. These groups form strong covalent bonds with the oxygen atoms, enhancing the material's structural integrity and stabilizing the redox processes. Such stability is crucial for maintaining the performance of SIBs over numerous charge–discharge cycles [96].

One of the primary advantages of polyanion cathodes is their enhanced safety. The strong covalent bonds and robust frameworks in these materials significantly mitigate the risk of oxygen release at high voltages, a phenomenon that can lead to thermal runaway in other types of cathodes. This characteristic makes polyanion cathodes particularly attractive for applications demanding high safety standards, such as stationary energy storage systems. This section delves into the key attributes of polyanion-type cathodes, with a focus on materials such as $Na_3V_2(PO_4)_3$ and $Na_2FePO_4F$, examining their performance and potential advantages for SIBs.

**$Na_3V_2(PO_4)_3$** stands out as a highly promising cathode material for SIBs due to its stable crystal structure and high operating voltage. This material exhibits several beneficial properties. Notably, $Na_3V_2(PO_4)_3$ operates at approximately 3.4 V, thanks to the strong covalent bonding within the $PO_4$ tetrahedra. This bonding enhances the structural stability of the material and ensures a stable voltage platform. The high operating voltage significantly boosts the overall energy density of the battery. Additionally, the stable crystal structure of $Na_3V_2(PO_4)_3$ minimizes volume changes during charge and discharge cycles, maintaining the mechanical integrity of the electrode and resulting in excellent cycle life. This stability is crucial for the long-term performance and reliability of the battery. Despite its advantages, $Na_3V_2(PO_4)_3$ has a lower theoretical capacity of about 120 mAh/g compared to some transition metal oxides. However, the high operating voltage and stable performance partially offset this limitation [97, 98].

**$Na_2FePO_4F$** combines the benefits of polyanions and fluorides, offering a high operating voltage and good capacity. It is thermally stable and safe for use in various applications. $Na_2FePO_4F$ operates at approximately 3.2 V, contributing to a favorable energy density. The inclusion of iron and fluorine in its structure enhances both stability and voltage. $Na_2FePO_4F$ has a capacity of about 120 mAh/g, which makes it competitive among polyanion cathode materials. The presence of fluoride ions helps maintain structural integrity during cycling, thereby improving the material's overall performance. Furthermore, $Na_2FePO_4F$ is known for its thermal stability, performing well at high temperatures and thus improving battery safety. This makes it particularly suitable for applications in high-temperature environments or devices subjected to significant thermal stress [99, 100].

In terms of electrochemical performance, polyanion cathode materials typically exhibit lower specific capacities compared to transition metal oxides. However, their high operating voltage and excellent stability contribute to their good overall performance. These properties make polyanion cathodes a compelling choice for applications where safety and long-term stability are prioritized over achieving the highest possible energy density [101, 102].

### 5.4.3 *Prussian blue analog cathodes*

Prussian blue analog (PBA) cathodes are highly promising for SIBs due to their distinctive open framework, which facilitates high sodium-ion mobility and supports relatively high operating voltages. The open framework significantly reduces the diffusion pathway for sodium ions, enabling rapid charge and discharge cycles. As a result, PBAs are particularly well suited for applications requiring high-power-density and fast-charging capabilities. The robust network formed by transition metals and cyanide groups in PBAs enhances electrochemical stability and facilitates efficient redox processes. Notable examples include Prussian blue (NaFe[Fe(CN)$_6$]) and Prussian white (Na$_2$Mn[Fe(CN)$_6$]), which strike a balance between high performance and cost-effectiveness, making them suitable for various energy storage applications [103, 104].

**Prussian blue**, illustrated in Figure 5.5, offers an operating voltage of approximately 3.4 V, owing to the strong Fe–C≡N–Fe bonds within its structure. This robust framework ensures minimal volume change during charge and discharge cycles, maintaining the mechanical integrity of the electrode and contributing to excellent cycle life. Additionally, Prussian blue is composed of inexpensive and readily available materials, making it a cost-effective choice for large-scale applications [105].

**Figure 5.5.** Prussian blue powder.

**Prussian white** features an open framework structure and achieves a high operating voltage of approximately 3.6 V, benefiting from the inclusion of manganese and iron, which enhance both its stability and voltage. With a capacity of about 150 mAh/g, Prussian white is competitive among PBA cathode materials. Its ability to perform well at high temperatures and improve battery safety makes it particularly suitable for high-temperature applications or devices experiencing significant thermal stress [106–108].

**Current challenges and disadvantages of Prussian white:** Despite its promising attributes, Prussian white faces several challenges that must be addressed for its practical application in SIBs. A major issue is its sensitivity to moisture. Prussian white is prone to degradation when exposed to moisture, which can significantly impact its electrochemical performance. The degradation mechanisms involve the formation of sodium hydroxide on the surface, leading to partial oxidation of $Fe^{2+}$ to $Fe^{3+}$. This process creates a passivating surface layer that inhibits ion transport and reduces capacity. Additionally, moisture can induce structural changes, causing irreversible capacity loss and reduced cycling stability. To improve the long-term performance of Prussian white cathodes, strategies such as applying surface coatings and maintaining controlled storage conditions are essential [109].

## 5.5 Anode Materials

Anode materials in SIBs are pivotal in determining the battery's overall performance, including its capacity, cycle life, and safety. SIB anodes can be broadly categorized based on their reaction mechanisms into three types: intercalation- or insertion-based materials, alloying compounds, and conversion-type materials. Each type of anode material offers distinct advantages and challenges, impacting the efficiency and application of SIBs.

### 5.5.1 Intercalation-type anode materials

Intercalation-type anode materials operate by enabling sodium ions to be intercalated or inserted into their structure during the charging process. These materials are known for their capacity to accommodate sodium ions

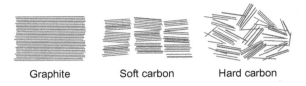

**Figure 5.6.** The morphologies of graphite, soft carbon, and hard carbon.

within their layered structures, which helps maintain stability and structural integrity of the anode throughout the cycling process. Examples of intercalation anodes include hard carbon (HC), soft carbon, and titanium-based materials (see Figure 5.6).

**Graphite**, commonly used as an anode material in LIBs, is unsuitable for SIBs due to the larger size of sodium ions compared to lithium ions. The interlayer spacing in graphite is not sufficient to accommodate sodium ions effectively, leading to poor intercalation and low capacity. This size mismatch results in limited sodium-ion storage within the graphite structure, rendering it inefficient as an anode material for SIBs [110].

In contrast, **HC** is one of the most promising intercalation anode materials for SIBs due to its high capacity and structural stability. HC features a disordered structure, comprising a mix of graphene sheets and nanopores that provide ample sites for sodium-ion storage. This disordered structure offers a large surface area and nanoporosity, making it well-suited for sodium-ion storage. Derived from various sources, including biomass, HC is also a sustainable option. Typically, HC anodes exhibit capacities ranging from 250 to 300 mAh/g and operate at low voltage plateaus (0.2–0.4 V vs. Na/Na$^+$), making them the current state of the art for SIBs [111].

**Soft carbon**, while offering lower capacity compared to HC, provides advantages in terms of initial coulombic efficiency and stability. Characterized by its layered structure, soft carbon facilitates sodium ion intercalation between the graphene layers. These materials typically exhibit capacities in the range of 200–250 mAh/g and operate at slightly higher voltage plateaus compared to HC, ranging from 0.1 to 1.0 V vs. Na/Na$^+$. The higher initial coulombic efficiency of soft carbon is attributed to fewer irreversible reactions during the first cycle. This characteristic enhances the overall efficiency and performance of the battery [112].

**Advantages and challenges:** Intercalation-type anodes hold great potential for SIBs, including relatively low charge/discharge voltage plateaus, good cycle life, and structural stability. However, challenges persist, such as optimizing specific capacity and improving initial Coulombic efficiency. Research efforts are ongoing to develop new intercalation materials and improve existing ones to further boost the performance and commercial viability of SIBs.

## 5.5.2 Alloying compounds

Alloying compounds are a prominent category of anode materials for SIBs due to their high theoretical capacities. These materials form alloys with sodium during the charging process, allowing them to store a large number of sodium ions. However, this alloying process involves significant volume changes, which pose challenges for maintaining the structural integrity and cycle life of the anode.

**Tin (Sn)** and **antimony (Sb)** are among the most studied alloying anode materials for SIBs. These materials can form various sodium-rich phases, such as $Na_{15}Sn_4$ and $Na_3Sb$, contributing to their high capacities. For instance, tin anodes can deliver theoretical capacities of up to 847 mAh/g, while antimony anodes can achieve around 660 mAh/g [113]. Despite these high capacities, the substantial volume expansion (up to 300% for Sn) during sodiation leads to mechanical stress and potential cracking of the anode material, which can degrade performance over multiple cycles [114].

**Composite and hybrid materials:** To address the issues of volume expansion and mechanical stress, researchers are developing composite and hybrid materials. These materials combine alloying compounds with carbon or other matrix materials to buffer volume changes and enhance structural stability of the anode. For example, composites such as Sn–C and Sb–C integrate tin or antimony with a carbon matrix, which helps to accommodate the volume changes and improve both the cycle life and rate capability of the anode [115].

**Advantages and challenges:** Alloying anodes offer high theoretical capacities, making them attractive for achieving high energy densities in SIBs. However, the significant volume changes associated with the alloying process pose substantial challenges for maintaining structural integrity and cycle stability. Therefore, developing strategies to mitigate these volume changes, such as using composite materials and optimizing electrode architectures, is crucial for the practical application of alloying anodes in SIBs.

### 5.5.3 *Conversion-type materials and mixed alloying–conversion systems*

Conversion-type materials are significant anode materials for SIBs due to their ability to undergo conversion reactions during the charge and discharge processes. These reactions involve the formation of new phases and multiple electron transfer reactions, leading to high theoretical capacities. Conversion-type materials typically include non-metallic elements such as phosphorus (P), sulfur (S), and oxygen (O), as well as metal oxides and sulfides.

**Phosphorus** is another promising alloying anode material with a high theoretical capacity of around 2596 mAh/g. When fully sodiated, phosphorus can form $Na_3P$, which contributes to its high capacity. However, similar to other alloying materials, phosphorus suffers from substantial volume expansion (~490%) during the sodiation process, leading to severe mechanical degradation and capacity fading [116].

**Sulfur and selenium** have also been explored as alloying anodes for SIBs due to their ability to form high-capacity sodiated phases, such as $Na_2S$ and $Na_2Se$. Sulfur, for instance, has a theoretical capacity of approximately 1675 mAh/g. However, challenges associated with sulfur and selenium anodes include significant volume changes and the dissolution of polysulfides/polyselenides in the electrolyte, which can lead to capacity loss and poor cycling stability [117].

**Metal oxides and sulfides:** Metal oxides, such as $Fe_2O_3$ and $Co_3O_4$, and metal sulfides, such as $MoS_2$ and $FeS_2$, are prominent conversion-type anode materials. These materials offer high capacities due to their ability to undergo multi-electron reactions. For example, iron oxide ($Fe_2O_3$) has a theoretical capacity of about 1007 mAh/g [118]. During the conversion reaction, the metal oxide or sulfide is reduced to form sodium oxide or sulfide and the corresponding metal, thereby providing high energy density. However, this process can also result in substantial volume changes and the formation of new phases, which contribute to mechanical degradation and capacity fading over repeated cycles [119].

**Advantages and challenges:** The primary advantage of conversion-type materials is their potential for high specific capacities, making them suitable for applications requiring high energy densities. However, they face significant challenges, including volume expansion during cycling, which can cause mechanical stress and electrode pulverization. Additionally, the formation of new phases can reduce electrical conductivity and slow reaction kinetics. To address these issues, researchers are investigating various strategies, such as using nanoscale materials, composites, and conductive matrices, to enhance the performance and cycle life of conversion-type anodes [120].

**Mixed alloying–conversion systems** combine the benefits of both alloying and conversion reactions to achieve higher capacities and improved stability. These systems leverage the high capacity typical of conversion reactions while incorporating alloying elements to buffer volume changes and enhance structural integrity. An example of a mixed alloying–conversion system is tin sulfide ($SnS_2$), which undergoes both alloying and conversion reactions during sodiation. Initially, $SnS_2$ forms $Na_2S$ through a conversion reaction, followed by the formation of Na–Sn alloys, combining high capacity with better cycle stability compared to materials relying solely on alloying or conversion reactions [121]. Another example is phosphorus–carbon composites, where the carbon matrix accommodates the volume changes associated with the conversion reaction of phosphorus, thereby improving cycle life and rate capability [122].

Despite their potential, mixed alloying–conversion systems face challenges related to volume expansion and the formation of insulating phases. Optimizing the composition and structure of these systems is essential for maximizing their performance. Advanced material design approaches, such as the development of nanoscale composites and the incorporation of conductive additives, can help address these challenges and improve the electrochemical properties of mixed alloying-conversion anodes [123].

## 5.6 Conclusion

The transition toward SIBs offers a promising solution to the limitations and dependencies associated with the current dominance of LIBs. A major challenge in the battery industry is the reliance on costly and geopolitically constrained materials, which affects both the economic feasibility and strategic security of large-scale battery production.

**Breaking the dependency on GWh factories:** To overcome the "curse" of GWh factories and make smaller-scale, economically viable battery solutions possible, it is essential to focus on utilizing cheaper and more widely available materials. Sodium, as opposed to lithium, is abundant and can be sourced locally in many regions, including Europe. This shift reduces reliance on the procurement power and resources concentrated in a few countries, thereby enhancing supply chain security and economic resilience.

**Advantages for Europe:** The adoption of SIBs and the utilization of locally available materials such as HC and Prussian blue/white cathodes offer substantial strategic benefits for Europe, particularly Germany. By relying on these materials, Europe can decrease its dependence on imported lithium and cobalt, which are subject to market volatility and geopolitical risks. The abundance of sodium, coupled with the development of compatible anode and cathode materials, can foster a more self-sufficient and sustainable battery industry within the region.

**Versatility and market potential of SIBs:** SIBs exhibit significant versatility and market, making them appealing for a broad spectrum of applications:

- Battery Electric Vehicles (BEVs): SIBs can deliver the required energy density and cycle life for electric vehicles, presenting a cost-effective alternative to LIBs.
- Replacement for Lead–Acid (PbA) Starter Batteries: SIBs offer robust performance, enhanced safety, and cost advantages that make them well suited to replace traditional lead–acid starter batteries in automotive applications. This shift can improve vehicle performance while reducing environmental impact.
- Light Electric Vehicles (LEVs): The lower cost and sufficient energy density of SIBs can drive the adoption of electric bikes, scooters, and other LEVs.
- Stationary Energy Storage Systems (ESS): SIBs are ideal for grid storage and renewable energy integration due to their high safety and cost-effectiveness. Their attributes make them an excellent choice for large-scale energy storage solutions.
- Power Tools and Portable Electronics: SIBs can efficiently meet the high-power and fast-charging demands of portable devices and power tools.

In conclusion, the advancement and adoption of SIBs present a strategic opportunity to enhance the sustainability, affordability, and security of energy storage technologies. By focusing on locally available materials and expanding the range of applications for SIBs, regions like Europe can position themselves at the forefront of the next generation of battery technology. This shift not only ensures long-term energy security and economic resilience but also fosters the development of smaller, cost-effective battery solutions that can overcome the limitations of traditional GWh-scale production. Investing in SIB technology can pave the way for a more diverse and sustainable energy future, benefiting a wide array of applications, from transportation to stationary energy storage.

# References

[1] H. Kim, Sodium-ion battery: Can it compete with Li-ion? *ACS Materials Au*, 3(6), 571–575, 2023.

[2] Nagmani, D. Pahari, A. Tyagi *et al.*, Lithium-ion battery technologies for electric mobility – state-of-the-art scenario, *ARAI Journal of Mobility Technology*, 2(2), 233–248, 2022.

[3] Z. Chen, A. Yildizbasi, Y. Wang *et al.*, Safety concerns for the management of end-of-life lithium-ion batteries, *Global Challenges*, 6(12), 2200049, 2022.

[4] T. Joshi, S. Azam, D. Juarez-Robles *et al.*, Safety and quality issues of counterfeit lithium-ion cells, *ACS Energy Letters*, 8(6), 2831–2839, 2023.

[5] K. Chayambuka, G. Mulder, D. L. Danilov *et al.*, From Li-ion batteries toward Na-ion chemistries: Challenges and opportunities, *Advanced Energy Materials*, 10(38), 2020.

[6] J.-N. Yang, H. Tian, K.-X. Wang *et al.*, 10 Years of frontiers in materials: Interface engineering for aqueous zinc-ion batteries, *Frontiers in Materials*, 11, 2024.

[7] A. Das, N. T. Balakrishnan, P. Sreeram *et al.*, Prospects for magnesium ion batteries: A compreshensive materials review, *Coordination Chemistry Reviews*, 502, 215593, 2024.

[8] J. Tu, W.-L. Song, H. Lei *et al.*, Nonaqueous rechargeable aluminum batteries: Progresses, challenges, and perspectives, *Chemical Reviews*, 121(8), 4903–4961, 2021.

[9] S. K. Das, S. Mahapatra, and H. Lahan, Aluminium-ion batteries: Developments and challenges, *Journal of Materials Chemistry A*, 5(14), 6347–6367, 2017.

[10] Q. Shao, S. Zhu, and J. Chen, A review on lithium-sulfur batteries: Challenge, development, and perspective, *Nano Research*, 16(6), 8097–8138, 2023.

[11] Y. Yi, F. Hai, J. Guo *et al.*, Progress and prospect of practical lithium-sulfur batteries based on solid-phase conversion, *Batteries*, 9(1), 27, 2023.

[12] Y. Liang, B. Zhang, Y. Shi *et al.*, Research on wide-temperature rechargeable sodium-sulfur batteries: Features, challenges and solutions, *Materials*, 16(12), 2023.

[13] L. Yaqoob, T. Noor, and N. Iqbal, An overview of metal-air batteries, current progress, and future perspectives, *Journal of Energy Storage*, 56, 106075, 2022.

[14] A. G. Olabi, E. T. Sayed, T. Wilberforce *et al.*, Metal-air batteries – A review, *Energies*, 14(21), 7373, 2021.

[15] I. Iwakiri, T. Antunes, H. Almeida *et al.*, Redox flow batteries: Materials, design and prospects, *Energies*, 14(18), 5643, 2021.

[16] Y. Wu and R. Holze, *Electrochemical Energy Conversion and Storage*, Wiley-VCH, Weinheim, 2022.

[17] T. Hosaka, K. Kubota, A. S. Hameed *et al.*, Research development on K-ion batteries, *Chemical Reviews*, 120(14), 6358–6466, 2020.

[18] Y. Liu and R. Holze, Metal-ion batteries, *Encyclopedia*, 2(3), 1611–1623, 2022.

[19] M. Sotoudeh, S. Baumgart, M. Dillenz *et al.*, Ion mobility in crystalline battery materials, *Advanced Energy Materials*, 14(4), 2024.

[20] S. Lei, Z. Zeng, M. Liu *et al.*, Balanced solvation/de-solvation of electrolyte facilitates Li-ion intercalation for fast charging and low-temperature Li-ion batteries, *Nano Energy*, 98, 107265, 2022.

[21] A. G. Hernandez, S. Cooper-Searle, A. C. Skelton *et al.*, Leveraging material efficiency as an energy and climate instrument for heavy industries in the EU, *Energy Policy*, 120, 533–549, 2018.

[22] J. B. Dunn, L. Gaines, J. C. Kelly *et al.*, The significance of Li-ion batteries in electric vehicle life-cycle energy and emissions and recycling's role in its reduction, *Energy & Environmental Science*, 8(1), 158–168, 2015.

[23] J. B. Dunn, L. Gaines, J. Sullivan *et al.*, Impact of recycling on cradle-to-gate energy consumption and greenhouse gas emissions of automotive lithium-ion batteries, *Environmental Science & Technology*, 46(22), 12704–12710, 2012.

[24] Y. Gao, Q. Yu, H. Yang *et al.*, The enormous potential of sodium/potassium-ion batteries as the mainstream energy storage technology for large-scale commercial applications, *Advanced materials (Deerfield Beach, Fla.)*, e2405989, 2024.

[25] B. Tang, L. Shan, S. Liang *et al.*, Issues and opportunities facing aqueous zinc-ion batteries, *Energy & Environmental Science*, 12(11), 3288–3304, 2019.

[26] K. W. Leong, W. Pan, X. Yi *et al.*, Next-generation magnesium-ion batteries: The quasi-solid-state approach to multivalent metal ion storage.

[27] J. Klemens, A.-K. Wurba, D. Burger *et al.*, Challenges and opportunities for large-scale electrode processing for sodium-ion and lithium-ion battery, *Batteries & Supercaps*, 6(11), 2023.

[28] A. G. Olabi, M. A. Allam, M. A. Abdelkareem *et al.*, Redox flow batteries: Recent development in main components, emerging technologies, diagnostic techniques, large-scale applications, and challenges and barriers, *Batteries*, 9(8), 409, 2023.

[29] S. Qiao, Q. Zhou, M. Ma *et al.*, Advanced anode materials for rechargeable sodium-ion batteries, *ACS nano*, 17(12), 11220–11252, 2023.

[30] X. Bai, N. Wu, G. Yu *et al.*, Recent advances in anode materials for sodium-ion batteries, *Inorganics*, 11(7), 289, 2023.

[31] H. A. Karahan Toprakci and O. Toprakci, Recent advances in new-generation electrolytes for sodium-ion batteries, *Energies*, 16(7), 3169, 2023.

[32] T. L. Kulova and A. M. Skundin, Electrode/electrolyte interphases of sodium-ion batteries, *Energies*, 15(22), 8615, 2022.

[33] F. A. Soto, Y. Ma, J. M. La Martinez de Hoz *et al.*, Formation and growth mechanisms of solid-electrolyte interphase layers in rechargeable batteries, *Chemistry of Materials*, 27(23), 7990–8000, 2015.

[34] A. A. Savina and A. M. Abakumov, Benchmarking the electrochemical parameters of the $LiNi0.8Mn0.1Co0.1O2$ positive electrode material for Li-ion batteries, *Heliyon*, 9(12), e21881, 2023.

[35] M. Lecompte, J. Bernard, E. Calas *et al.*, Experimental assessment of high-energy high nickel-content NMC lithium-ion battery cycle life at cold temperatures, *Journal of Energy Storage*, 94, 112443, 2024.

[36] S. Xu, H. Chen, C. Li *et al.*, A new high-performance $O3-NaNi0.3Fe0.2Mn0.5O2$ cathode material for sodium-ion batteries, *Ionics*, 29(5), 1873–1885, 2023.

[37] S. J. An, J. Li, C. Daniel *et al.*, The state of understanding of the lithium-ion-battery graphite solid electrolyte interphase (SEI) and its relationship to formation cycling, *Carbon*, 105, 52–76, 2016.

[38] L. Gao, J. Chen, Q. Chen *et al.*, The chemical evolution of solid electrolyte interface in sodium metal batteries, *Science Advances*, 8(6), eabm4606, 2022.

[39] J. Betz, G. Bieker, P. Meister *et al.*, Theoretical versus practical energy: a plea for more transparency in the energy calculation of different rechargeable battery systems, *Advanced Energy Materials*, 9(6), 2019.

[40] M. Chen, Y. Zhang, G. Xing *et al.*, Building high power density of sodium-ion batteries: Importance of multidimensional diffusion pathways in cathode materials, *Frontiers in Chemistry*, 8, 152, 2020.

[41] T. Hosaka, K. Kubota, A. S. Hameed *et al.*, Research development on K-ion batteries, *Chemical Reviews*, 120(14), 6358–6466, 2020.

[42] E. J. Kim, P. R. Kumar, Z. T. Gossage *et al.*, Active material and interphase structures governing performance in sodium and potassium ion batteries, *Chemical Science*, 13(21), 6121–6158, 2022.

[43] T. Akcay, M. Haeringer, K. Pfeifer *et al.*, Degradation mechanisms of positive electrode materials for sodium-ion batteries, *ECS Meeting Abstracts*, MA2023-01(5), 911, 2023.

[44] A. Mukhopadhyay and B. W. Sheldon, Deformation and stress in electrode materials for Li-ion batteries, *Progress in Materials Science*, 63, 58–116, 2014.

[45] F. A. Soto, Y. Ma, J. M. La Martinez de Hoz *et al.*, Formation and growth mechanisms of solid-electrolyte interphase layers in rechargeable batteries, *Chemistry of Materials*, 27(23), 7990–8000, 2015.

[46] J. S. Edge, S. O'Kane, R. Prosser *et al.*, Lithium ion battery degradation: what you need to know, *Physical Chemistry Chemical Physics: PCCP*, 23(14), 8200–8221, 2021.

[47] J. Weaving, J. Robinson, D. Ledwoch *et al.*, Sodium-ion batteries: aging, degradation, failure mechanisms and safety, in M.-M. Titirici, P. Adelhelm, and Y.-S. Hu (Eds.), *Sodium-Ion Batteries*, pp. 501–530, Wiley, 2022.

[48] Y. Yu, *Sodium-Ion Batteries: Energy Storage Materials and Technologies*, Wiley-VCH, Weinheim, 2022.

[49] S. Guo, Y. Sun, J. Yi *et al.*, Understanding sodium-ion diffusion in layered P2 and P3 oxides via experiments and first-principles calculations: A bridge between crystal structure and electrochemical performance, *NPG Asia Materials*, 8(4), e266-e266, 2016.

[50] L. Xiao, F. Ji, J. Zhang *et al.*, Doping regulation in polyanionic compounds for advanced sodium-ion batteries, *Small (Weinheim an der Bergstrasse, Germany)*, 19(1), e2205732, 2023.

[51] M. Chen, J. Xiao, W. Hua *et al.*, A cation and anion dual doping strategy for the elevation of titanium redox potential for high-power sodium-ion batteries, *Angewandte Chemie (International ed. in English)*, 59(29), 12076–12083, 2020.

[52] H. Jiang, G. Qian, R. Liu *et al.*, Effects of elemental doping on phase transitions of manganese-based layered oxides for sodium-ion batteries, *Science China Materials*, 66(12), 4542–4549, 2023.

[53] L. Ran, A. Baktash, M. Li *et al.*, Sc, Ge co-doping NASICON boosts solid-state sodium ion batteries' performance, *Energy Storage Materials*, 40, 282–291, 2021.

[54] A. Yang, X. Huang, C. Luo *et al.*, High-rate-capacity cathode based on Zn-doped and carbonized polyacrylonitrile-coated Na4MnV(PO4)3 for sodium-ion batteries, *ACS Applied Materials & Interfaces*, 15(18), 22132–22141, 2023.

The Role of Raw Materials in Enhancing or Limiting Energy Density    181

[55] Y. Chen, Q. Shi, S. Zhao *et al.*, Investigation on the air stability of P2-layered transition metal oxides by Nb doping in sodium ion batteries, *Batteries*, 9(3), 183, 2023.

[56] J. Yan, H. Li, K. Wang *et al.*, Ultrahigh phosphorus doping of carbon for high-rate sodium ion batteries anode, *Advanced Energy Materials*, 11(21), 2021.

[57] F. Zhang, J. Liao, L. Xu *et al.*, Stabilizing P2-type Ni-Mn oxides as high-voltage cathodes by a doping-integrated coating strategy based on zinc for sodium-ion batteries, *ACS Applied Materials & Interfaces*, 13(34), 40695–40704, 2021.

[58] H. Fang, H. Ji, J. Zhai *et al.*, Mitigating Jahn-Teller effect in layered cathode material via interstitial doping for high-performance sodium-ion batteries, *Small (Weinheim an der Bergstrasse, Germany)*, 19(35), e2301360, 2023.

[59] H. Ma, J. Bai, P. Wang *et al.*, Double-carbon-layer coated Na 4 MnV (PO 4)3 towards high-performance sodium-ion full batteries, *Advanced Materials Interfaces*, 9(30), 2022.

[60] L. Zhu, B. Yin, Y. Zhang *et al.*, A multifunctional coating on sulfur-containing carbon-based anode for high-performance sodium-ion batteries, *Molecules (Basel, Switzerland)*, 28(8), 2023.

[61] R. D. Chakraborty, M. Bhar, S. Bhowmik *et al.*, Synergistic effect of 3D electrode architecture and in situ carbon coating on the electrochemical performance of SnO 2 anodes for sodium-ion batteries, *Journal of the Electrochemical Society*, 171(4), 40521, 2024.

[62] X. Chen, C. Cheng, M. Ding *et al.*, Elucidating the redox behavior in different P-type layered oxides for sodium-ion batteries, *ACS Applied Materials & Interfaces*, 12(39), 43665–43673, 2020.

[63] T. Song, W. Yao, P. Kiadkhunthod *et al.*, A low-cost and environmentally friendly mixed polyanionic cathode for sodium-ion storage, *Angewandte Chemie (International ed. in English)*, 59(2), 740–745, 2020.

[64] Q. Liu, W. Zheng, G. Liu *et al.*, Realizing high-performance cathodes with cationic and anionic redox reactions in high-sodium-content P2-type oxides for sodium-ion batteries, *ACS Applied Materials & Interfaces*, 2023.

[65] Y. Fang, X.-Y. Yu, and X. W. Lou, Nanostructured electrode materials for advanced sodium-ion batteries, *Matter*, 1(1), 90–114, 2019.

[66] J. B. Goodenough and Y. Kim, Challenges for rechargeable Li batteries, *Chemistry of Materials*, 22(3), 587–603, 2010.

[67] L. Zhao, L. Kang, S. Yao *et al.*, Nanostructured NaTi2(PO4)3@C particles embedded into nitrogen-doped graphene sheets as novel anode materials for sodium-ion batteries, *Solid State Ionics*, 335, 47–52, 2019.

[68] L. Chen, B. Kishore, and E. Kendrick, Nanostructured materials for sodium-ion batteries, in *Nanomaterials for Electrochemical Energy*

[69] F. Li and Z. Zhou, Micro/nanostructured materials for sodium ion batteries and capacitors, *Small (Weinheim an der Bergstrasse, Germany)*, 14(6), 2018.

[70] Q. Wang, C. Zhao, Y. Lu *et al.*, Advanced nanostructured anode materials for sodium-ion batteries, *Small (Weinheim an der Bergstrasse, Germany)*, 13(42), 2017.

[71] C. Zhu, P. Kopold, W. Li *et al.*, Engineering nanostructured electrode materials for high performance sodium ion batteries: A case study of a 3D porous interconnected WS 2/C nanocomposite, *Journal of Materials Chemistry A*, 3(41), 20487–20493, 2015.

[72] H. Zhang, I. Hasa, and S. Passerini, Sodium-ion batteries: Beyond insertion for Na-ion batteries: Nanostructured alloying and conversion anode materials (Adv. Energy Mater. 17/2018), *Advanced Energy Materials*, 8(17), 2018.

[73] G.-L. Xu, R. Amine, Y.-F. Xu *et al.*, Insights into the structural effects of layered cathode materials for high voltage sodium-ion batteries, *Energy & Environmental Science*, 10(7), 1677–1693, 2017.

[74] C. Shi, L. Wang, X. Chen *et al.*, Challenges of layer-structured cathodes for sodium-ion batteries, *Nanoscale Horizons*, 7(4), 338–351, 2022.

[75] Z. Yan, L. Liu, J. Tan *et al.*, One-pot synthesis of bicrystalline titanium dioxide spheres with a core–shell structure as anode materials for lithium and sodium ion batteries, *Journal of Power Sources*, 269, 37–45, 2014.

[76] B. Cheng, S. Zhang, H. Zhuo *et al.*, Crystalline iceplant-like nano-NaVPO4F@graphene as an intercalation-type anode material for sodium-ion batteries, *Chemical Communications (Cambridge, England)*, 56(16), 2479–2482, 2020.

[77] C. Li, R. Zang, P. Li *et al.*, High crystalline Prussian white nanocubes as a promising cathode for sodium-ion batteries, *Chemistry, an Asian journal*, 13(3), 342–349, 2018.

[78] M. Chen, W. Hua, J. Xiao *et al.*, NASICON-type air-stable and all-climate cathode for sodium-ion batteries with low cost and high-power density, *Nature Communications*, 10(1), 1480, 2019.

[79] Z.-Y. Gu, Z.-H. Sun, J.-Z. Guo *et al.*, High-rate and long-cycle cathode for sodium-ion batteries: Enhanced electrode stability and kinetics via binder adjustment, *ACS Applied Materials & Interfaces*, 12(42), 47580–47589, 2020.

[80] P. Verma and S. Puravankara, Jute-fiber precursor-derived low-cost sustainable hard carbon with varying micro/mesoporosity and distinct storage mechanisms for sodium-ion and potassium-ion batteries, *Langmuir: The ACS Journal of Surfaces and Colloids*, 38(50), 15703–15713, 2022.

[81] S. Trivedi, V. Pamidi, S. P. Bautista *et al.*, Water-soluble inorganic binders for lithium-ion and sodium-ion batteries, *Advanced Energy Materials*, 14(9), 2024.

[82] Z. Mao, R. Wang, B. He *et al.*, Cross-linked sodium alginate as a multifunctional binder to achieve high-rate and long-cycle stability for sodium-ion batteries, *Small (Weinheim an der Bergstrasse, Germany)*, 19(11), e2207224, 2023.

[83] X. Liu, Y. Cao, and J. Sun, Defect engineering in Prussian blue analogs for high-performance sodium-ion batteries, *Advanced Energy Materials*, 12(46), 2022. https://onlinelibrary.wiley.com/doi/epdf/10.1002/aenm.202202532.

[84] D. Zhang, Y. Shao, J. Wang *et al.*, Cobalt-mediated defect engineering endows high reversible amorphous VS4 anode for advanced sodium-ion storage, *Small (Weinheim an der Bergstrasse, Germany)*, 20(27), e2309901, 2024.

[85] M. A. Muhammad, D. Pan, Y. Liu *et al.*, N-doped 3D carbon encapsulating nickel selenide nanoarchitecture with cation defect engineering: An ultrafast and long-life anode for sodium-ion batteries, *Journal of Colloid and Interface Science*, 670, 191–203, 2024.

[86] L. Zhong, H. Chen, W. Xie *et al.*, Intercalation and defect engineering of layered MnPS3 for greatly enhanced capacity and stability in sodium-ion batteries, *Chemical Engineering Journal*, 481, 148370, 2024.

[87] Y. Yu, Transition metal oxide cathodes for sodium-ion batteries, in Y. Yu (Ed.), *Sodium-Ion Batteries*, pp. 41–78, Wiley, 2022.

[88] Z. Liu, X. Xu, S. Ji *et al.*, Recent progress of P2-type layered transition-metal oxide cathodes for sodium-ion batteries, *Chemistry (Weinheim an der Bergstrasse, Germany)*, 26(35), 7747–7766, 2020.

[89] S. Li, Y. Sun, Y. Pang *et al.*, Recent developments of layered transition metal oxide cathodes for sodium-ion batteries toward desired high performance, *Asia-Pacific Journal of Chemical Engineering*, 17(4), 2022. https://onlinelibrary.wiley.com/doi/epdf/10.1002/apj.2762.

[90] L. Gan, X.-G. Yuan, J.-J. Han *et al.*, Highly symmetrical six-transition metal ring units promising high air-stability of layered oxide cathodes for sodium-ion batteries, *Advanced Functional Materials*, 33(7), 2023. https://advanced.onlinelibrary.wiley.com/doi/epdf/10.1002/adfm.202209026.

[91] L. Yao, P. Zou, C. Wang *et al.*, High-entropy and superstructure-stabilized layered oxide cathodes for sodium-ion batteries, *Advanced Energy Materials*, 12(41), 2022. https://onlinelibrary.wiley.com/doi/epdf/10.1002/aenm.202201989.

[92] Y. Wang, Y. Wang, Y. Xing *et al.*, Entropy modulation strategy of P2-type layered transition metal oxide cathodes for sodium-ion batteries with a

high performance, *Journal of Materials Chemistry A*, 11(37), 19955–19964, 2023.

[93] H. Gao, J. Zeng, Z. Sun *et al.*, Advances in layered transition metal oxide cathodes for sodium-ion batteries, *Materials Today Energy*, 42, 101551, 2024.

[94] Q. Wang, S. Chu, and S. Guo, Progress on multiphase layered transition metal oxide cathodes of sodium ion batteries, *Chinese Chemical Letters*, 31(9), 2167–2176, 2020.

[95] J. Xiao, Y. Xiao, J. Li *et al.*, Advanced nanoengineering strategies endow high-performance layered transition-metal oxide cathodes for sodium-ion batteries, *SmartMat*, 4(5), 2023.

[96] S.-P. Guo, J.-C. Li, Q.-T. Xu *et al.*, Recent achievements on polyanion-type compounds for sodium-ion batteries: syntheses, crystal chemistry and electrochemical performance, *Journal of Power Sources*, 361, 285–299, 2017.

[97] T. Jin, H. Li, K. Zhu *et al.*, Polyanion-type cathode materials for sodium-ion batteries, *Chemical Society Reviews*, 49(8), 2342–2377, 2020.

[98] Q. Ni, Y. Bai, F. Wu *et al.*, Polyanion-type electrode materials for sodium-ion batteries, *Advanced Science (Weinheim, Baden-Wurttemberg, Germany)*, 4(3), 600275, 2017.

[99] Y. Gao, H. Zhang, X.-H. Liu *et al.*, Low-cost polyanion-type sulfate cathode for sodium-ion battery, *Advanced Energy Materials*, 11(42), 2021. https://onlinelibrary.wiley.com/doi/epdf/10.1002/aenm.202101751.

[100] Z. Song, R. Liu, W.-D. Liu *et al.*, Low-cost polyanion-type cathode materials for sodium-ion battery, *Advanced Energy and Sustainability Research*, 4(11), 2023. https://onlinelibrary.wiley.com/doi/epdf/10.1002/aesr.202300102.

[101] X. Wu, A. Liu, and S. Lu, Enhancing voltage output in polyanion-type cathode materials for sodium ion batteries, *Batteries & Supercaps*, 2024.

[102] M. Li, Engineering routes towards practical sodium-ion batteries: case studies from oxides to polyanion compounds, *ECS Meeting Abstracts*, MA2023-01(5), 913, 2023.

[103] Y. Yu, Prussian blue analogue cathodes for sodium-ion batteries, in Y. Yu (Ed.), *Sodium-Ion Batteries*, pp. 137–159, Wiley, 2022.

[104] T. Yuan, Y. Chen, X. Gao *et al.*, Research progress of Prussian blue and its analogs as high-performance cathode nanomaterials for sodium-ion batteries, *Small Methods*, e2301372, 2023. https://onlinelibrary.wiley.com/doi/epdf/10.1002/smtd.202301372.

[105] S. Syed Mohd Fadzil, H. J. Woo, A. D. Azzahari *et al.*, Sodium-rich prussian blue analogue coated by poly(3,4-ethylenedioxythiophene) polystyrene sulfonate as superior cathode for sodium-ion batteries, *Materials Today Chemistry*, 30, 101540, 2023.

[106] H. Zhang, Y. Gao, J. Peng *et al.*, Prussian blue analogues with optimized crystal plane orientation and low crystal defects toward 450 Wh kg-1 alkali-ion batteries, *Angewandte Chemie (International ed. in English)*, 62(27), e202303953, 2023.

[107] Y. Lu, S. Vail, X. Zhao *et al.*, High performance sodium-ion batteries based on Prussian white electrode, *ECS Meeting Abstracts*, MA2016-03(2), 335, 2016.

[108] C. Q. X. Lim and Z.-K. Tan, Prussian white with near-maximum specific capacity in sodium-ion batteries, *ACS Applied Energy Materials*, 4(6), 6214–6220, 2021.

[109] D. O. Ojwang, M. Svensson, C. Njel *et al.*, Moisture-driven degradation pathways in Prussian white cathode material for sodium-ion batteries, *ACS Applied Materials & Interfaces*, 13(8), 10054–10063, 2021.

[110] T. Zhang, C. Li, F. Wang *et al.*, Recent advances in carbon anodes for sodium-ion batteries, *Chemical Record* (New York, N.Y.), 22(10), e202200083, 2022.

[111] Y. Yu, Intercalation-type anode materials for sodium-ion batteries, in Y. Yu (Ed.), *Sodium-Ion Batteries*, pp. 203–243, Wiley, 2022.

[112] Z. Xiao, Q. Li, Y. Yang *et al.*, Intercalation and alloying anode materials for rechargeable Li/Na batteries, in Y. Yang, R. Fu, and H. Huo (Eds.), *NMR and MRI of Electrochemical Energy Storage Materials and Devices*, pp. 253–280, The Royal Society of Chemistry, 2021.

[113] M. Lao, Y. Zhang, W. Luo *et al.*, Alloy-based anode materials toward advanced sodium-ion batteries, *Advanced Materials* (Deerfield Beach, Fla.), 29(48), 2017.

[114] C. Hu, X. Hou, Z. Bai *et al.*, Promises and challenges of Sn-based anodes for sodium-ion batteries, *Chinese Journal of Chemistry*, 39(10), 2931–2942, 2021.

[115] H. Xie, W. P. Kalisvaart, B. C. Olsen *et al.*, Sn–Bi–Sb alloys as anode materials for sodium ion batteries, *Journal of Materials Chemistry A*, 5(20), 9661–9670, 2017.

[116] A. Rehman, S. Saleem, S. Ali *et al.*, Recent advances in alloying anode materials for sodium-ion batteries: Material design and prospects, *Energy Materials*, 4(6), 2024.

[117] L. Fang, N. Bahlawane, W. Sun *et al.*, Conversion-alloying anode materials for sodium ion batteries, *Small (Weinheim an der Bergstrasse, Germany)*, 17(37), e2101137, 2021.

[118] H. Liu, S. Luo, S. Yan, Q. Wang, D. Hu, Y. Wang *et al.*, High-performance $\alpha$-Fe2O3/C composite anodes for lithium-ion batteries synthesized by hydrothermal carbonization glucose method used pickled iron oxide red as raw material, *Composites Part B: Engineering*, 164, 576–582, 2019.

[119] J. Zhang, K. Song, L. Mi *et al.*, Bimetal synergistic effect induced high reversibility of conversion-type Ni@NiCo2S4 as a free-standing anode for sodium ion batteries, *The Journal of Physical Chemistry Letters*, 11(4), 1435–1442, 2020.

[120] S. Wang, Y. Dong, F. Cao *et al.*, Conversion-type MnO nanorods as a surprisingly stable anode framework for sodium-ion batteries, *Advanced Functional Materials*, 30(19), 2020.

[121] L. Wu, H. Lu, L. Xiao *et al.*, A tin(ii) sulfide–carbon anode material based on combined conversion and alloying reactions for sodium-ion batteries, *Journal of Materials Chemistry A*, 2(39), 16424–16428, 2014.

[122] Z. Huang, H. Hou, C. Wang *et al.*, Molybdenum phosphide: A conversion-type anode for ultralong-life sodium-ion batteries, *Chemistry of Materials*, 29(17), 7313–7322, 2017.

[123] S. Sarkar and S. C. Peter, An overview on Sb-based intermetallics and alloys for sodium-ion batteries: Trends, challenges and future prospects from material synthesis to battery performance, *Journal of Materials Chemistry A*, 9(9), 5164–5196, 2021.

© 2025 World Scientific Publishing Company
https://doi.org/10.9789811282058_0006

# Chapter 6

# Remaining Energy Densities at the Battery Systems Level

**Julian Joël Grimm[*] and Kai Peter Birke[†]**

*Fraunhofer IPA, Nobelstrasse 12, Stuttgart, Germany*

*[*]julian.grimm@ipa.fraunhofer.de*

*[†]kai.peter.birke@ipa.fraunhofer.de*

This chapter provides a comprehensive overview of current and future trends in determining the remaining energy densities at the battery system level for automotive applications, highlighting the main technological developments and challenges. The focus is on the three basic cell concepts: cylindrical, prismatic pouch, and prismatic hardcase. Each cell concept has specific advantages and disadvantages. Regardless of the battery cells used, elements for cooling and controlling as well as the mechanical stability required for vehicle operation must also be considered at the battery system level. These components occupy additional space, increase the system's mass, and thus affect overall efficiency. The integration of the batteries in vehicles is similar across all concepts due to size and weight considerations, typically located in the underbody. Consequently, these effects lead to similar achievable energy densities for automotive batteries across all variants. Considering the development of energy densities for lithium-ion batteries, an almost horizontal asymptote of energy densities is to be expected. It is anticipated that energy densities at the battery system level will experience minor increases before converging to their maximum potential. One of the most interesting new developments is the blade concept, in which the battery becomes self-supporting. This innovation is currently only feasible with the hardcase concept. Based on the conclusions of the previous chapters, the cylindrical cell concept is currently deemed the most promising option for automotive applications.

## 6.1 Cell Formats for Automotive Applications

In automotive applications, battery systems power the vehicle's drive system using the individual battery cells with their limited energy content integrated into battery modules. The goal is to construct battery systems with a nominal voltage of 400–800 V and an energy capacity of approximately 40–100 kWh. The requirements for building battery systems include optimizing the available installation space to achieve high gravimetric and volumetric energy density at the system level while also ensuring safe operation of the battery through electrical, thermal, and mechanical components. The three cell concepts, cylindrical, prismatic pouch, and prismatic hardcase, serve as the baseline for constructing battery systems, as introduced in previous chapters. Currently, all three cell concepts are employed in the construction of BEVs, with each OEM typically using a uniform cell concept.

Figure 6.1 exemplarily shows a battery system by Tesla for their Model S 2021 built using cylindrical cells. Figure 6.2 illustrates a system of the Volkswagen ID3 2020 based on pouch cells, also used by manufacturers such as Mercedes-Benz, Hyundai, and BYD. Figure 6.3 depicts a system of the BMW iX3 2021 based on prismatic cells.

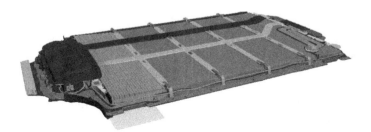

**Figure 6.1.** Battery system with cyclindrical cells, exemplified by the Tesla Model S 2021 (schematic view).

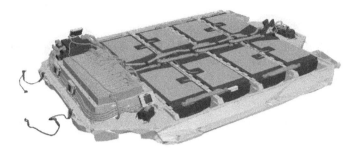

**Figure 6.2.** Battery system with prismatic hardcase cells, exemplified by the BMW iX3 2021 (schematic view).

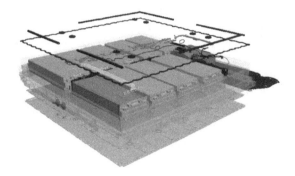

**Figure 6.3.** Battery system with pouch cells, exemplified by the Volkswagen ID3 2020 (schematic view).

Currently, each cell concept has its own advantages and disadvantages, as described in Chapter 4. Therefore, all three cell concepts are fundamentally suitable for use in battery systems for automotive applications.

## 6.2 Automotive Requirements for Battery Systems

To examine the use of different cell concepts in automotive battery systems, it is important to understand the basic structure and components of a battery system. Beyond the battery cells integrated in cell modules, the main components include the thermal system, the electrical system with its electrical and electronical components, control units such as the battery management system (BMS), busbars and cables, as well as the mechanical structure, which includes housing and insulation. These components collectively occupy space and contribute to an increase in the mass of the battery system. As non-active elements, they impact the energy density of the battery system.

### 6.2.1 *Battery cells and modules for energy content*

For the cell-to-module design of the three cell concepts, there are both fundamental challenges and advantages. The primary challenge with cylindrical cells is that many small individual cells need to be connected. However, module sizes can be adapted to fit specific requirements, and defects in individual cells have a relatively minor impact on the overall system due to their smaller size. In contrast, pouch cells present challenges in module construction because the module size is significantly influenced by the cell size. These cells require clamping and a stable housing. On the positive side, fewer connection elements are needed compared to cylindrical cells, owing to the larger cell size. When constructing cell-to-module systems using prismatic hardcase cells, the challenge lies in the module size being significantly influenced by the cell size, and the weight of the cell housings adds substantial mass to the overall system. However, similar to pouch cells, prismatic hardcase cells also benefit from fewer required connection elements due to their larger size. Additionally, their cubic shape facilitates straightforward packaging.

### 6.2.2 *Thermal system for cooling*

The primary function of the thermal system is to supply and dissipate thermal energy to optimize the battery's operating temperature. The goal is to achieve the most uniform temperature distribution possible throughout the entire battery system. Key requirements for the thermal system include preventing "thermal runaway" due to high temperatures (>60°C), avoiding high temperatures to reduce (calendar) aging, heating the battery to enhance performance (as higher temperature leads to lower resistance), and enabling fast charging while reducing cyclic aging. The design of the thermal system fundamentally depends on the cell concept, as different designs exhibit varying thermal conductivity properties. Generally, thermal conductivity is better along the electrode but poorer across it.

Therefore, each of the three cell concepts used in automotive applications requires specific cooling solutions and strategies.

For battery systems composed of cylindrical cells, the numerous small individual cells result in a high surface-to-capacity ratio, facilitating effective cooling through the surface alone. This is typically achieved using cooling spirals placed between the cells, as illustrated in Figure 6.4 for the Tesla Model S, which utilizes cylindrical cells.

In battery systems using pouch cells, there is considerable thermal resistance from the heat source within the cell, passing through the pouch film, module housing, cooling plate, and ultimately to the cooling medium. This creates an inert system with a high time constant. Cooling systems for pouch cells often involve mounting the modules onto a cooling plate. Alternatively, cooling plates may be designed to specifically target the contacts of the negative poles, where heat input is the highest. Figure 6.5 illustrates the thermal cooling system implemented in the Volkswagen ID3, which employs pouch cells.

**Figure 6.4.** Thermal system of a battery system with prismatic hardcase cells, exemplified by the BMW iX3 2021 (schematic view).

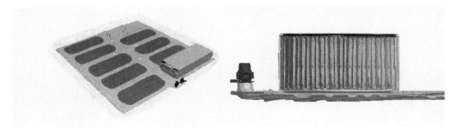

**Figure 6.5.** Thermal system of a battery system with pouch cells, exemplified by the Volkswagen ID3 2020 (schematic view).

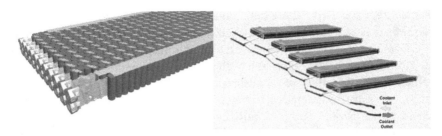

**Figure 6.6.** Thermal system of a battery system with cylindrical cells, exemplified by the Tesla Model S 2021 (schematic view).

The thermal system of prismatic hardcase cells is typically implemented using a cooling plate as the base of the module. This design ensures effective cooling but also requires an additional connection for the cooling system, alongside the electrical connections. Figure 6.6 depicts the thermal cooling system used in the BMW iX3, which is based on prismatic hardcase cells.

Other cooling concepts in battery systems include passive cooling, air cooling, and immersion cooling. In passive cooling, the thermal mass of the battery system is utilized solely for cooling purposes. Air cooling relies on the direct flow of air around the cells, eliminating the need for additional components such as cooling plates. However, this method requires a high volume flow due to the low specific heat capacity of air, as well as air treatment to prevent contamination within the battery system. The immersion cooling concept involves flooding the battery system housing with dielectric cooling fluid, which also negates the need for additional components such as cooling plates. While these systems can achieve very high performance, their implementation is very complex. Furthermore, the dielectric cooling fluids used (usually oils) can be very aggressive and corrosive, necessitating significant efforts to ensure proper sealing.

## 6.2.3 *Electrical system for controlling*

The primary function of the electrical system is to ensure and control a safe operation of the battery. This system includes components such as the BMS with master and module controllers, measurement technology for current, voltage, and temperature, fuse and switching elements, as well as busbars and cables. There are no inherent differences in the electrical system based on cell concepts. However, the integration of additional power electronic components into the battery housing may sometimes be necessary. Typically, all electrical and electronic components are either concentrated in a specific area or part of the battery system or distributed throughout the system. This distribution can lead to variations in component accessibility. Figures 6.7–6.9 illustrate the electrical systems for the different battery configurations previously discussed.

Regardless of the specific grouping and placement of the electrical and electronic components, it is evident that these components require additional space and contribute to the overall weight of the battery system in all variants.

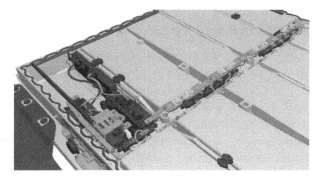

**Figure 6.7.** Electrical system of a battery system with prismatic hardcase cells, exemplified by the BMW iX3 2021 (schematic view).

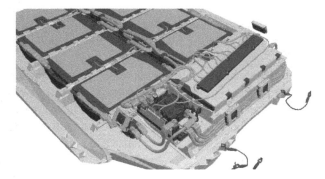

**Figure 6.8.** Electrical system of a battery system with pouch cells, exemplified by the Volkswagen ID3 2020 (schematic view).

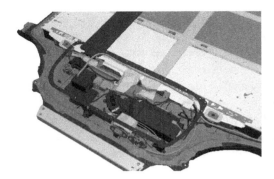

**Figure 6.9.** Electrical system of a battery system with cylindrical cells, exemplified by the Tesla Model S 2021 (schematic view).

### 6.2.4 *Mechanical structure for stability*

In addition to the cells and the thermal and electrical systems, the mechanical structure of the battery system also includes additional passive components.

For battery systems using cylindrical cells, stability is primarily achieved through the inherent robustness of the cylindrical cells themselves, supplemented by an outer flange, a sheet metal structure, and additional support points. Further mechanical modules are generally not required. Figure 6.10 illustrates the mechanical structure of the Tesla Model S, which is based on cylindrical cells.

For pouch cells, additional elements are required due to their mechanical instability. The mechanical structure consists of a solid and rigid module along with a system housing, which are necessary for clamping the individual cells and ensuring mechanical stability. Figure 6.11 depicts the mechanical structure for the Volkswagen ID3, which utilizes pouch cells.

Prismatic hardcase cells, on the other hand, require only a frame for the modules due to their inherent mechanical stability. However, like pouch cells, they also need a solid and rigid system housing made of individual profiles. Figure 6.12 illustrates the mechanical structure for the BMW iX3, which is based on prismatic hardcase cells.

The generation of mechanical stability through the required structural components necessitates additional installation space and contributes to increased weight. Due to the lower stability of individual pouch cells, the

*Remaining Energy Densities at the Battery Systems Level* 195

**Figure 6.10.** Mechanical structure of a battery system with prismatic hardcase cells, exemplified by the BMWiX3 2021 (schematic view).

**Figure 6.11.** Mechanical structure of a battery system with pouch cells, exemplified by the Volkswagen ID3 2020 (schematic view).

**Figure 6.12.** Mechanical structure of a battery system with cylindrical cells, exemplified by the Tesla Model S 2021 (schematic view).

effort required for the implementation is greater compared to cylindrical and prismatic hardcase cells.

## 6.2.5 *Integration of the battery system in BEV*

When considering the integration of the battery system with all components in the vehicle, the approach is consistent for all cell concepts. In the past, battery systems were installed in available spaces or in areas made available by removing unnecessary components in combustion-engine vehicles (see Figure 6.13). Today, however, due to their size and weight, battery systems are exclusively placed in the vehicle's underbody, regardless of the cell concept (see Figure 6.14). The size of the battery system is tailored to fit the vehicle frame, and its installation is aligned with the vehicle's structural framework.

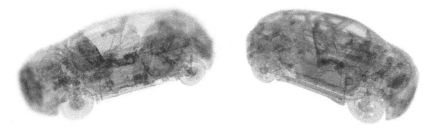

**Figure 6.13.** Integration of a battery system in a BEV where space is available, exemplified by the eGolf 2014 and BMW i3 2014 (schematic view).

**Figure 6.14.** Integration of a battery system in the vehicle's underbody, exemplified by the Volkswagen ID3 2020 (schematic view).

Regarding the installation situation, there are differences in height buildup among the various cell concepts. Cylindrical cells have a lower height buildup due to their smaller dimensions, while prismatic hardcase cells exhibit the highest height buildup (see Figures 6.15–6.17).

**Figure 6.15.** Integration height of a battery system with cylindrical cells in a BEV, exemplified by the Tesla Model S (schematic view).

**Figure 6.16.** Integration height of a battery system with pouch cells in a BEV, exemplified by the Volkswagen ID3 (schematic view).

**Figure 6.17.** Integration height of a battery system with prismatic hardcase cells in a BEV, exemplified by the BMW iX3 (schematic view).

## 6.3. Remaining Energy Densities at the Battery System Level

After examining the main components and differences at the battery system level arising from the design and assembly of various cell concepts, the question of the achievable remaining energy density at the battery system level for the different cell concepts becomes pertinent. The target specifications for a battery system can be directly derived from the requirements of the application, specifically BEVs. The sub-objectives include cost (particularly TCO), range (both WLTP and real-world), performance data (continuous power, peak power, and peak torque), safety, fast-charging capability, and lifespan (calendar and cyclic life). The energy density of a battery system is implicitly relevant to several of these objectives. Since the main components – comprising the thermal system, electrical system, and mechanical structure – are similar across all cell concepts, the achievable energy densities for automotive battery systems are also comparable. This is illustrated by the examples listed in Table 6.1.

Based on the energy densities achievable today, it is expected that further developments will converge toward a horizontal asymptote in the near future. Rather than experiencing a breakthrough, there will likely be a gradual evolution in energy density at the battery system level. The maximum potential of lithium-ion technology appears increasingly capped, with approximately 250 Wh/kg at the system level and 350 Wh/kg at the cell level serving as useful rule-of-thumb values for future expectations.

**Table 6.1.** Examples for the remaining energy densities at battery cell (C), module (M) and system (S) level

| Example (cell concept) | Cell connection | Mass (kg) | Capacity (kWh) | Voltage (V) | Energy density C/M/S (Wh/kg) | Energy density C/M/S (Wh/l) |
|---|---|---|---|---|---|---|
| Tesla Model S 2021 (cylindrical) | 110S72P | 552 | 100 | 407 | 257 / 230 / 173 | 720 / 515 / 255 |
| Volkswagen ID3 2020 (pouch) | 108S2P | 377 | 58 | 396 | 270 / 225 / 167 | 693 / 495 / 249 |
| BMW iX3 2021 (prismatic hardcase) | 94S2P | 513 | 80 | 345 | 245 / 222 / 156 | 498 / 450 / 181 |

## 6.4 Upcoming Developments

This section explores emerging trends in the key components of battery systems.

In the electrical system, there is a growing trend toward 800 V battery systems, particularly for fast-charging applications. These systems offer advantages in charging infrastructure by reducing the required electrical load and current. However, adapting 800 V battery systems to existing 400 V charging stations requires adjustments. Solutions include using motor inductance, as seen with the Hyundai Ioniq 5, or splitting the battery into two 400 V stacks, as demonstrated by the Tesla Cybertruck.

In the mechanical system, the current trend favors cylindrical and pouch cells over prismatic hardcase cells, primarily due to the weight disadvantage posed by the heavy housing of prismatic hardcase cells. For example, BMW plans to transition from prismatic hardcase cells to cylindrical cells in their new class of vehicles. Additionally, there are ongoing efforts to develop Cell2Vehicle architectures instead of traditional Cell2Module2System2Vehicle solutions to optimize weight. In this context, the reusability and recyclability of battery components in terms of a circular economy should be critically examined.

In the thermal system, efforts are underway to develop modular solutions for battery cooling systems, allowing for more flexible placement of individual components within the vehicle.

In the software domain, there is a trend toward creating more advanced BMS functionalities. This includes state-of-health (SoH) determination through cycling, current pulses, and electrochemical impedance spectroscopy (EIS) measurements, as well as expanded EIS functionalities for determining temperature, state-of-charge (SoC), and SoH. Additionally, the trend toward vehicle centralization and the software-defined vehicle concept is prompting the outsourcing of BMS functions or components of the battery module controller to a central controller.

One of the most intriguing recent developments in battery technology is the blade concept. In this design, the battery system features long and thin battery cells that resemble blades. These cells are directly integrated into the battery system, eliminating the need for modules, which saves space and enhances efficiency. The design facilitates a denser packing of cells, enabling greater energy storage within the same space. However, this self-supportive structure is only applicable for hardcase cells, rendering pouch cells unsuitable for this development.

Safety considerations often counteract the benefits of larger cell capacities, particularly with regard to thermal runaway and thermal propagation in large pouch cells, assuming the same cell chemistry. In contrast, cylindrical cells face limitations in capacity due to their more complex tab configurations. This may create a "sweet spot" where cylindrical 46XXX cells emerge as an optimal choice in the future.

In summary, considering the remaining energy density at the battery system level and upcoming developments, cylindrical cells currently represent the most promising cell concept for battery systems in automotive applications.

© 2025 World Scientific Publishing Company
https://doi.org/10.9789811282058_0007

# Chapter 7

# Achievable Progress in Battery Energy Densities

**Yucheng Luo[*] and Kai Peter Birke[†]**

*Fraunhofer IPA, Nobelstrasse 12, Stuttgart, Germany*

*[*]yucheng.luo@ipa.fraunhofer.de*

*[†]kai.peter.birke@ipa.fraunhofer.de*

## 7.1 Energy Density of Batteries

As global demands for sustainable energy solutions intensify, advancements in battery technology are pivotal. In the evolving landscape of energy storage technology, the role of battery energy density emerges as a critical factor in the operational efficiency and viability of both mobile and stationary applications. This chapter aims to delve into the inherent trade-offs between energy density and battery lifespan, as well as the challenges faced by energy density in the battery degradation process. It seeks to provide a comprehensive overview of the current state of battery technology and its development toward greater efficiency and sustainability.

### 7.1.1 *Capacity, energy density, and power density*

Among various electrochemical energy storage technologies, lithium-ion batteries (LIBs) stand out due to their superior energy density and relatively low production and operational costs, making them widely adopted. The critical parameters for LIBs include energy density, power density, cycle life, cost per kilowatt-hour, and safety. In the context of electric vehicles (EVs), driving range (energy) and charging speed (power) are two crucial characteristics.

High-energy batteries typically feature thicker electrodes with a higher volumetric fraction of active materials, thereby increasing energy content and driving range. In contrast, high-power batteries usually have thinner electrodes to reduce internal resistance, with a higher volumetric fraction of electrolytes and conductive additives, enhancing power capability and acceleration. Achieving both high energy and high power density in a single cell is challenging due to these differing requirements [1, 2].

Apart from electrode thickness, the energy density of a battery significantly depends on the chemical materials used, particularly the cathode materials. The primary cathode materials for LIBs include lithium cobalt oxide ($LiCoO_2$), lithium iron phosphate ($LiFePO_4$), lithium manganese oxide ($LiMn_2O_4$), and various ternary lithium-ion compounds (NMC and NCA). The choice of cathode material is pivotal, as it determines the upper limit of the cell's energy density. Safety, energy density, and power density form the basis for selecting cathode materials for EVs, each having a theoretical energy density that influences overall performance [3].

Therefore, in battery design and material selection, a trade-off between energy and power is often necessary, typically represented by the power-to-energy (P/E) ratio.

As shown in Table 7.1, studies have compared the characteristics of three mainstream LIBs. It is evident that, although LFP has a lower nominal capacity (Ah) and nominal voltage (V) compared to NMC and NCA, it boasts a longer cycle life until 80% of its usable nominal capacity is reached [4].

**Table 7.1.** Battery characteristics by common lithium-ion battery chemistries [5–7].

| Li-ion battery chemistry | Cell-level specific energy [Wh/kg] | Nominal voltage [V] | Nominal capacity [Ah] | Cycle life [Cycles] | Calendar life [Years] |
|---|---|---|---|---|---|
| NMC–Graphite | 140–200 | 3.7 | 2.5–3.5 | 2000+ | 8–10 |
| NCA–Graphite | 200–250 | 3.6 | 2.9–3.6 | 2000+ | 8–10 |
| LFP–Graphite | 90–140 | 3.2 | 1.5–2.5 | 3000+ | 8–12 |
| LFP–LTO | ≤80 | 2.7 | 0.8–1.2 | 5000+ | 10+ |
| LCO–Graphite | 140–190 | 3.7 | 2.8 | 1000 | 4–7 |

Capacity loss was evaluated in terms of the change in state of health (SoH) with an increasing number of equivalent full cycles (EFCs), where one EFC is measured based on the cell's initial capacity:

$$N_{EFC} = \frac{\text{Total discharge throughput}}{\text{Nominal capacity}}.$$

According to several studies, as demonstrated in Figure 7.1, the LFP cells exhibit substantially longer cycle lifespans than other chemistries under similar conditions: over 3000 EFCs compared to 800–1000 EFCs for LCO cells and around 2800 EFCs for NMC cells; besides, 250–1500 EFCs for NCA according to an existing study [4].

Furthermore, while LFP cells demonstrated the highest cycle lifetime across all conditions, this performance gap diminished when cells were compared according to discharge energy throughput. This value is calculated by summing the energy discharged from the battery during each cycle. When considering the lower capacity and voltage of LFP batteries,

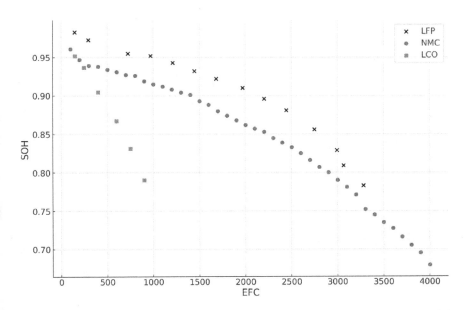

**Figure 7.1.** Capacity fade vs. equivalent full cycle (EFC). All batteries were cycled under the same conditions (charging and discharging rate of 1 C and within a SoC window of 0–100%) [9–11].

the performance differences among the three chemistries become less pronounced [4].

In conclusion, optimizing the chemical composition of batteries is essential for balancing energy density, power density, and cycle life. Understanding the trade-offs and interactions between different materials and designs enables the development of batteries that meet specific application needs, particularly in the rapidly evolving field of EVs.

### 7.1.2 *Aging mechanisms*

#### 7.1.2.1 *Overview of aging mechanisms*

The aging mechanisms of LIBs are complex and multifaceted, impacting battery performance and lifespan through various electrochemical reactions, material degradation, and structural changes. Understanding these mechanisms is crucial for improving battery design and extending operational life. As illustrated in Figure 7.2, the aging processes in LIBs can be categorized into three main aspects: effects of electrochemical reactions, impact of material degradation, and influence of structural changes.

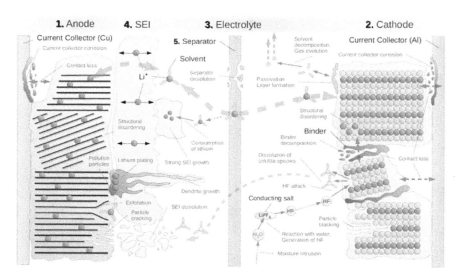

**Figure 7.2.** Degradation mechanisms in Li-ion batteries [5].

## (a) **Effects of electrochemical reactions**

The electrochemical reactions within a LIB play a significant role in its aging process. These reactions can lead to several detrimental effects:

1. *Current Collector Corrosion (Anode and Cathode Current Collector Corrosion)*: As shown in Figure 7.2, at both the anode and cathode, the current collectors (copper and aluminum) undergo corrosion, leading to contact loss and current collector malfunction.

   In the electrolyte, current collector materials may dissolve or react, producing by-products that further affect battery performance.

2. *Lithium Plating*: Lithium plating occurs at the anode, where lithium ions cannot intercalate into the graphite structure and instead deposit on the anode surface as metallic lithium.

   This plating can lead to the formation of lithium dendrites, increasing the risk of short circuits.

3. *Solvent Decomposition*: Solvents in the electrolyte decompose under high-voltage or high-temperature conditions, generating gases and increasing internal pressure within the battery.

   These by-products can react with other electrolyte components, leading to further degradation of the electrolyte.

## (b) **Impact of material degradation**

Material degradation is another critical aspect of battery aging, involving changes in the physical and chemical properties of battery materials:

1. *Particle Cracking*: In electrode materials (both cathode and anode), repeated charge and discharge cycles induce mechanical stress and cracking of active material particles.

   Cracking increases the surface area of electrode materials, accelerating side reactions with the electrolyte.

2. *Exfoliation*: Exfoliation of anode materials, especially graphite electrodes, results in contact loss of active material, reducing battery capacity and cycle life.

   Cathode materials may also undergo similar exfoliation or structural degradation.

3. *Dissolution of Soluble Species*: Transition metals (e.g., cobalt, nickel, and manganese) in the cathode dissolve into the electrolyte during charge and discharge, leading to material loss and performance degradation.

These dissolved metal ions can redeposit at the anode, further affecting anode performance.

(c) **Influence of structural changes**

Structural changes within the battery components are a third major contributor to battery aging. These changes can affect the overall stability and performance of the battery:

1.  *Structural Disordering*: In electrode materials, repeated lithium-ion insertion and extraction cause crystal structure disordering and expansion, reducing material stability and conductivity.

    Cathode materials such as NMC (nickel cobalt manganese oxides) are prone to crystal structure changes at high voltages, leading to capacity decay.
2.  *SEI Growth and Dissolution*: On the anode surface, the growth of the solid–electrolyte interphase (SEI) layer protects the anode; however, an excessively thick SEI layer increases internal resistance, reducing battery efficiency.

    The dissolution and regrowth of the SEI layer consume lithium ions, leading to irreversible capacity loss.
3.  *Binder Decomposition*: The binder in electrodes decomposes under high temperatures or electrochemical conditions, compromising the mechanical integrity of electrode materials.

    Binder degradation causes active material detachment and contact loss, further reducing battery performance.

The above list includes some of the most commonly reported degradation mechanisms in Li-ion batteries. Additionally, there are other mechanisms such as cathode electrolyte interphase (CEI) and cation mixing. Overall, these various aging mechanisms collectively contribute to the degradation of the battery cell. Understanding these mechanisms is essential for developing strategies to mitigate aging and enhance the durability of LIBs.

### 7.1.2.2 *Battery degradation and its energy density*

Energy density, defined as the amount of energy stored per unit volume or mass, is a crucial metric for evaluating battery performance. A higher energy density indicates that a battery can store more energy within a given volume or weight, making it an essential parameter for energy storage systems. The impact of aging mechanisms, specifically capacity fade, increase in internal resistance, and thermal management issues, on the energy density of battery energy storage systems (BESS) is significant and multifaceted.

Capacity fade refers to the gradual reduction in a battery's usable capacity over repeated charge and discharge cycles. As the battery ages, internal processes such as mechanical wear, heat stress, and chemical and electrochemical reactions lead to a decrease in the total amount of energy that can be stored and delivered. This reduction directly impacts the energy density, as the available energy per unit volume or mass diminishes. The degradation of active materials and the formation of passivation layers (e.g., SEI) on the electrodes further contribute to capacity loss. Consequently, the energy density declines as the battery can no longer store the same amount of energy as it could when new. Furthermore, capacity fade results in the need for more frequent charging in practical use, indirectly affecting the effective lifespan of the battery and its energy density.

The increase in internal resistance is another aging mechanism that affects energy density. As a battery undergoes cycling, internal resistance tends to rise due to factors such as electrolyte decomposition, SEI layer growth, and electrode material degradation. Higher internal resistance results in greater energy losses during charge and discharge cycles, as more energy is dissipated as heat rather than stored or delivered as usable electrical energy. This inefficiency reduces the effective energy density of the battery. Moreover, the increased resistance causes a drop in the terminal voltage under load, which limits the usable capacity and further decreases the energy density.

Thermal management is critical for maintaining battery performance and safety. Ineffective thermal management can lead to elevated temperatures, which accelerate aging processes such as electrolyte degradation, electrode material deterioration, and SEI layer growth. High temperatures can exacerbate capacity fade and increase internal resistance, thereby reducing energy density. Additionally, thermal gradients within the battery pack can cause uneven aging, leading to imbalances in cell performance and further diminishing the overall energy density. Effective thermal management strategies are essential to mitigate these effects and maintain higher energy density over the battery's lifespan.

### 7.1.3 *Factors influencing energy density*

Energy density in LIBs is influenced by a multitude of factors that encompass environmental conditions, load conditions, and operational time. Understanding these factors is critical for optimizing battery performance and longevity. In this section, we delve into how temperature, state of charge (SoC), C-rate, depth of discharge (DoD), and time contribute to various aging mechanisms, ultimately affecting the energy density of LIBs. Specifically, we focus on degradation mechanisms at both the component and cell levels, including SEI formation, lithium plating, electrolyte decomposition, gas formation, transition metal dissolution, cation mixing, and particle cracking.

1. **SEI Formation (conditions: high SoC, high temperature, high C-rate, and high DoD):**
   The SEI layer forms on the anode surface during the initial cycles of a battery. While it passivates the anode and prevents continuous electrolyte decomposition, its growth is significantly accelerated under conditions of high SoC, high temperature, high C-rate, and high DoD. As the SEI layer thickens, it increases the internal resistance of the battery, which reduces the overall energy density. Over time, the continuous formation and repair of the SEI layer consume active lithium and electrolyte, further diminishing capacity and energy density.

2. **Lithium Plating (conditions: high SoC, low temperature, high C-rate, and high DoD):**
   Lithium plating occurs when lithium ions deposit as metallic lithium on the anode surface instead of intercalating into the anode material. This phenomenon is particularly pronounced under conditions of high SoC, low temperature, high C-rate, and high DoD. Lithium plating can lead to the loss of active lithium and create dendrites that may cause internal short circuits, severely compromising battery safety and performance and reducing the overall energy density.

3. **Electrolyte Decomposition (conditions: high SoC, high temperature, and high DoD):**
   The electrolyte in a LIB is susceptible to decomposition, especially at high SoC, high temperatures, and high DoD. Decomposition products can form gases, increasing internal pressure and potentially causing

cell swelling or rupture. Additionally, the consumption of the electrolyte reduces its availability for ion transport, negatively affecting the battery's performance and energy density.

4. **Gas Formation (conditions: high SoC, high temperature, high C-rate, and high DoD):**
   Gas formation results from the decomposition of the electrolyte and other side reactions. High SoC, high temperatures, high C-rates, and high DoD exacerbate these processes, leading to an increase in internal pressure. Gas formation can cause cell swelling, increase internal resistance, and pose safety risks, ultimately reducing the effective energy density of the battery.

5. **Transition Metal Dissolution (conditions: high SoC and high temperature):**
   Transition metals from the cathode can dissolve into the electrolyte under high SoC and high temperatures. These dissolved metals can migrate to the anode, where they precipitate and increase resistance. This process degrades the cathode structure and diminishes the battery's overall capacity and energy density.

6. **Cation Mixing (conditions: high SoC, high temperature, and high DoD):**
   Cation mixing involves the disordering of lithium and transition metal ions in the cathode structure. High SoC, high temperatures, and high DoD conditions facilitate this process, which disrupts the cathode's crystalline structure and reduces its ability to store lithium ions effectively, thereby lowering the energy density.

7. **Particle Cracking (conditions: high C-rate and high DoD):**
   Rapid charging and discharging (high C-rate) and deep discharges (high DoD) induce significant mechanical stress on the electrode particles, leading to cracking. These cracks can isolate active material, increase resistance, and create pathways for side reactions, resulting in lower capacity and reduced energy density.

   In summary, the energy density of LIBs is influenced by a complex interplay of environmental factors, load conditions, and operational time. High temperatures, high SoC, high C-rates, and high DoD are particularly detrimental, accelerating various aging mechanisms, such as SEI and CEI formation, lithium plating, electrolyte decomposition, gas formation, transition metal dissolution, cation mixing, and particle cracking.

### 7.1.4 *The impact of operating conditions on battery degradation mechanisms and optimization strategies*

Battery aging is an extremely complex dynamic evolution process, where different conditions trigger various aging mechanisms that can interact with each other, making the prediction of battery aging highly challenging. This explains why the actual lifespan of batteries often deviates from the standard expected lifespan provided by manufacturers. Even when considering only LIBs, the optimal operating conditions vary depending on the specific anode and cathode materials used. Furthermore, other battery technologies, such as lead–acid, solid-state, fuel cell, and lithium–metal batteries, exhibit significant differences in their chemical properties, making them suitable for different application scenarios. Therefore, it is impractical to compare the lifespans of different battery technologies under a single set of environmental conditions.

A more feasible approach is to discuss the most suitable battery technology and its corresponding optimal operating conditions within a specific application scenario (controlling certain environmental variables). Currently, due to factors such as technological maturity, industrial scalability, and cost, LIBs are widely used in electrical complexes and systems, including as traction batteries in EVs. Their usage rate far exceeds that of other battery technologies. Although the manufacturing cost of LIBs has continuously decreased over the past decade (see Figure 7.3), the cost of batteries still accounts for approximately 40% of the total production cost of EVs. Therefore, optimizing the load strategies and environmental conditions for LIBs to maximize their lifespan is a major focus of current battery research.

As previously mentioned, environmental factors and operational loads have a decisive impact on battery lifespan. The following sections describe how these factors affect battery degradation, whether in a linear, exponential, or other form. Understanding these effects is crucial because, in many cases, practical needs require trade-offs in operational strategies. Only by fully understanding the significance and interrelationships of each factor can appropriate adjustments and choices be made.

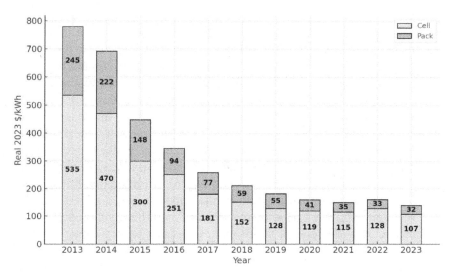

**Figure 7.3.** Volume-weighted average lithium-ion battery pack and cell price split from 2013–2023 [12].

Specifically, temperature, charge/discharge rates, depth of discharge, and charging cut-off voltage are critical factors influencing battery lifespan. For instance, operating at excessively high or low temperatures can significantly accelerate battery aging, potentially leading to electrolyte decomposition and instability of the SEI layer. High-rate charging and discharging can exacerbate chemical and mechanical stress within the battery, accelerating the degradation of active materials. Deep discharges and high charging cut-off voltages can cause structural changes in electrode materials, increasing the likelihood of side reactions. The combined effects of these factors result in a complex nonlinear degradation process.

In practical applications, considering the diverse requirements, trade-offs between battery lifespan, performance, and safety may be necessary. For example, in EV applications, users may prioritize range and charging speed, whereas in energy storage systems, cycle life and stability may be more critical. Therefore, researching and optimizing battery operational strategies for different application scenarios is crucial for enhancing overall battery efficiency.

## 7.1.4.1 *Depth of discharge*

Controlling the DoD of batteries is an effective method to maintain stable charge and discharge performance and ensure battery longevity by not utilizing a portion of the battery's capacity during charging and discharging. Although increasing the DoD range can easily extend the driving range per charge, this practice also brings significant adverse effects [13].

When the DoD range is increased, the cathode material undergoes excessive delithiation, triggering a series of undesirable side reactions. The irreversible structural changes in the cathode material can significantly accelerate battery performance degradation and lead to the decomposition of the active cathode material [14–16]. Additionally, lithium ions lost from the cathode can deposit on the anode active material, causing side reactions at the anode–electrolyte interface. These side reactions are generally considered one of the primary causes of battery capacity decay [17, 18].

Specifically, excessive delithiation of the cathode material not only affects its structural integrity but also exacerbates electrolyte decomposition and instability of the interface layer, thereby reducing the battery's cycle life and performance. Lithium deposition on the anode can form lithium dendrites, further increasing the risk of short circuits and thermal runaway. In the worst-case scenario, the battery may experience overcharging during cell balancing, leading to thermal runaway and potential safety hazards [19].

To optimize battery lifespan and performance, it is crucial to precisely control the DoD range and avoid overutilization of battery capacity. Research and development of more stable cathode and anode materials, as well as improvements in battery management systems (BMS) to better monitor and regulate the charging and discharging process, are essential directions in current battery technology development. These measures can extend battery life while ensuring safety and reliability, thus meeting the demands of modern EVs and energy storage systems for efficient and durable battery technologies.

Theoretically, a smaller DoD – implying minimal charge–discharge cycles – is beneficial for prolonging battery life. One study suggests that deep cycling induces disproportionate or exponential degradation, as illustrated in Figure 7.4 [20, 21]. Contrarily, another study indicates that the dependency of battery aging on cycle depth exhibits a more linear

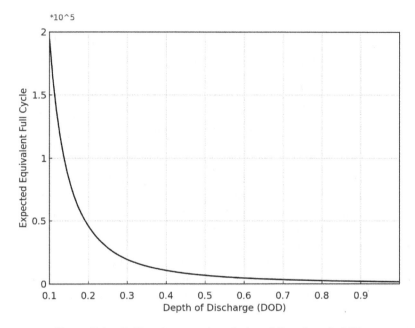

**Figure 7.4.** DoD and expected equivalent full cycles of a LIB.

trend rather than an exponential one [22]. The impact of DoD on battery degradation may vary depending on its chemical composition, but overall, a larger DoD cycle accelerates battery degradation.

However, in practical applications, we must balance operational feasibility and convenience. Taking into account factors such as charge–discharge energy and Coulombic efficiency, it has been demonstrated that a DoD range of 60–70% achieves optimal battery performance [13]. This range strikes a balance between extending battery life and maintaining practical usability.

To achieve an optimal DoD, it is essential to consider the trade-offs between battery longevity and the need for sufficient energy throughput. A moderate DoD range provides a compromise that maximizes battery life while delivering adequate energy storage capacity. This ensures that batteries can support high energy demands without significantly sacrificing longevity, making them a practical choice for various applications, including EVs and stationary energy storage systems.

### 7.1.4.2 State of charge

When considering calendar aging, lower SoC levels are generally less detrimental compared to higher SoC levels [23]. Within the entire voltage range, aging factors of capacity and resistance exhibit a linear trend, as shown in Figure 7.5 [22]. For instance, in 18650 batteries with NMC cathode chemistry, a higher storage SoC corresponds to a higher voltage, leading to greater capacity loss and increased resistance. The primary driver of capacity degradation during storage is identified as the low anode potential. In high SoC states, where the lithiation rate of the graphite anode exceeds 50%, the low anode potential accelerates the loss of cyclable lithium, thereby altering electrode balance. Especially at 100% SoC, NMC and NCA batteries may experience additional aging effects due to high battery voltage-induced side reactions, such as electrolyte oxidation and transition metal dissolution [24].

However, when factoring in cycling aging, the situation changes. Cycling at moderate voltages results in the least aging. Both capacity loss and resistance increase are minimized when cycling occurs between 45% and 55% SoC. Cycling at either lower or higher voltages increases the rate of cyclic aging [22]. This U-shaped function of average voltage, with a minimum around 3.7 V, has been corroborated by another study [25],

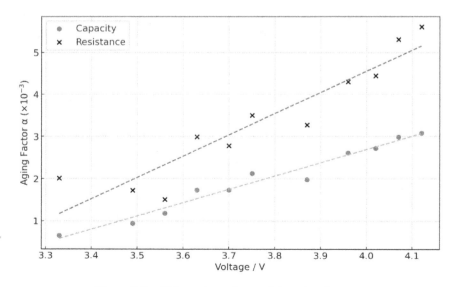

**Figure 7.5.** Voltage dependency of the aging factor $\alpha$.

which demonstrates that the impact of cycling at intermediate SoC levels is less severe, both in terms of capacity loss and resistance increase. Maintaining a moderate average SoC level helps prolong battery lifespan by minimizing high/low voltage-induced side reactions while balancing the practical requirements of energy storage and retrieval.

### 7.1.4.3 Temperature

When batteries operate and are stored above room temperature, the temperature dependence is well established in the literature, with the Arrhenius equation commonly used to describe the linear correlation of the logarithm of aging factors (such as capacity and resistance) with temperature (as shown in Figure 7.6) [22]:

$$k = Ae^{-E_a/RT},$$

where $k$ is the rate constant of the reaction, $A$ is the pre-exponential factor, which is a constant for each chemical reaction, $E_a$ is the activation energy for the reaction (in joules per mole or kilojoules per mole), $R$ is the universal gas constant, and $T$ is the absolute temperature.

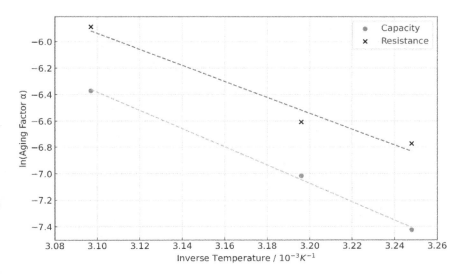

**Figure 7.6.** Arrhenius plot of aging factor $\alpha$ for both capacity and resistance over inverse temperature.

In contrast, operating and storing batteries below room temperature, especially in extreme cold, significantly shortens battery lifespan. Low temperatures increase the impedance of LIBs, resulting in power and energy loss [26]. Specific studies have indicated that the maximum SoC of a battery can drop by approximately 7–23% when the ambient temperature drops below 0°C, which is insufficient for many applications [27, 28]. The energy density decreases from 100 Wh/kg at 25°C to 5 Wh/kg at −40°C, and the power density falls from 800 to 10 Wh/kg [27, 29].

During charging and regenerative braking, low temperatures significantly increase the risk of lithium plating, leading to lithium loss and negatively impacting battery longevity [30, 31]. According to current research, the optimal temperature for battery performance is 25°C [25]. Temperatures above 25°C also shorten battery life but are not as detrimental as low temperatures (where the difference exceeds 200%). The most favorable temperature range for LIBs is between 25°C and 40°C.

The poor performance of LIBs at low temperatures can be attributed to several factors: (a) slow chemical reaction characteristics and charge transfer rates, resulting in poor kinetic performance and increased SEI resistance at low temperatures; (b) decreased electrolyte conductivity; (c) lithium dendrite deposition at the anode, which, after repeated cycling, leads to dendrite formation, causing short circuits and battery failure [27, 28, 32]. These factors pose significant challenges to the application of LIBs in low-temperature environments.

### 7.1.4.4 *Current rate*

Theoretically, increasing the C-rate accelerates battery aging. However, the C-rate doesn't play as important a role as the aforementioned factors. A study indicates that the average SoC exerts a more significant influence on battery degradation. Higher C-rates at lower SoC levels cause less aging compared to lower C-rates at higher SoC levels. Notably, even at a 4C discharge rate, performance remains favorable under low SoC conditions [23].

While high C-rates do contribute to battery aging, their more critical impact lies in heat generation. According to the power equation ($P = I^2R$), an increase in current leads to an exponential increase in thermal losses. This significantly affects battery efficiency and, in severe cases, can result in thermal runaway, posing substantial safety risks.

Understanding the influence of environmental and load factors on battery degradation is crucial. This knowledge enables the simulation and prediction of battery aging based on environmental and load conditions alone, without needing to delve deeply into the specific degradation mechanisms. This predictive capability is essential for advancing battery technology, ensuring safety, and maximizing the lifespan of battery systems in various applications.

# References

[1]   F. Naseri, C. Barbu, and T. Sarikurt, Optimal sizing of hybrid high-energy/high-power battery energy storage systems to improve battery cycle life and charging power in electric vehicle applications, *Journal of Energy Storage*, 55, 105768, 2022.

[2]   M. Quarti, A. Bayer, W. G. Bessler, Trade-off between energy density and fast-charge capability of lithium-ion batteries: A model-based design study of cells with thick electrodes, *Electrochemical Science Advances*, 3(1), 2023.

[3]   J. Wen, D. Zhao, and C. Zhang, An overview of electricity powered vehicles: Lithium-ion battery energy storage density and energy conversion efficiency, *Renewable Energy*, 162, 1629–1648, 2020.

[4]   Y. Preger, H. M. Barkholtz, A. Fresquez, *et al.* Degradation of commercial lithium-ion cells as a function of chemistry and cycling conditions, *Journal of the Electrochemical Society*, 167(12), 120532, 2020.

[5]   Christoph R. Birkl, Matthew R. Roberts, Euan McTurk, Peter G. Bruce, David A. Howey, Degradation diagnostics for lithium ion cells, *Journal of Power Sources*, 341, 373–386, 2017. https://doi.org/10.1016/j.jpowsour.2016.12.011. (https://www.sciencedirect.com/science/article/pii/S0378775316316998)

[6]   V. Vega-Garita, A. Hanif, N. Narayan *et al.*, Selecting a suitable battery technology for the photovoltaic battery integrated module, *Journal of Power Sources,* 438, 227011, 2019.

[7]   J. J. Lamb and B. G. Pollet, *Micro-Optics and Energy*, 2020.

[8]   J. Porzio and C. D. Scown, Life-cycle assessment considerations for batteries and battery materials. *Advanced Energy Materials,* 11(33), 2021.

[9]   Y. Gao, J. Jiang, C. Zhang *et al.*, Lithium-ion battery aging mechanisms and life model under different charging stresses. *Journal of Power Sources,* 356, 103–114, 2017.

[10]  E. Sarasketa-Zabala, I. Gandiaga, E. Martinez-Laserna *et al.*, Cycle ageing analysis of a LiFePO4/graphite cell with dynamic model

validations: Towards realistic lifetime predictions. *Journal of Power Sources*, 275, 573–587, 2015.

[11] F. Richter, P. J. Vie, S. Kjelstrup *et al.*, Measurements of ageing and thermal conductivity in a secondary NMC-hard carbon Li-ion battery and the impact on internal temperature profiles. *Electrochimica Acta*, 250, 228–237, 2017.

[12] https://about.bnef.com/blog/lithium-ion-battery-pack-prices-hit-record-low-of-139-kwh/

[13] S.-J. Park, Y.-W. Song, B.-S. Kang *et al.* Depth of discharge characteristics and control strategy to optimize electric vehicle battery life, *Journal of Energy Storage*, 59, 106477, 2023.

[14] Y. Zeng, K. Wu, D. Wang *et al.*, Overcharge investigation of lithium-ion polymer batteries, *Journal of Power Sources*, 160(2), 1302–1307, 2006.

[15] M. Ouyang, D. Ren, L. Lu *et al.* Overcharge-induced capacity fading analysis for large format lithium-ion batteries with LiNi1/3Co1/3Mn1/3O2+ Li Mn2O4 composite cathode, *Journal of Power Sources*, 279, 626–635, 2015.

[16] D. Belov and M.-H. Yang, Failure mechanism of Li-ion battery at overcharge conditions, *Journal of Solid State Electrochemistry*, 12(7–8), 885–894, 2008.

[17] Pankaj Arora *et al.* Capacity Fade Mechanisms and Side Reactions in Lithium-Ion Batteries, *Journal of The Electrochemical Society*, 145, 3647, 1998.

[18] L. Tan, L. Zhang, Q. Sun *et al.*, Capacity loss induced by lithium deposition at graphite anode for LiFePO4/graphite cell cycling at different temperatures, *Electrochimica Acta*, 111, 802–808, 2013.

[19] L. Huang, Z. Zhang, Z. Wang *et al.*, Thermal runaway behavior during overcharge for large-format Lithium-ion batteries with different packaging patterns, *Journal of Energy Storage*, 25, 100811, 2019.

[20] D. U. Sauer, Secondary batteries–lead–acid systems: Lifetime determining processes. In: *Encyclopedia of Electrochemical Power Sources*. Elsevier, pp. 805–815, 2009.

[21] J. Suh, S. Song, and G. Jang, Power imbalance-based droop control for vehicle to grid in primary frequency regulation, *IET Generation, Transmission & Distribution*, 16(17), 3374–3383, 2022.

[22] J. Schmalstieg, S. Kabitz, M. Ecker *et al.*, From accelerated aging tests to a lifetime prediction model: Analyzing lithium-ion batteries, pp. 1–12.

[23] E. Wikner and T. Thiringer, Extending battery lifetime by avoiding high SOC, *Applied Sciences*, 8(10), 1825, 2018.

[24] P. Keil, S. F. Schuster, J. Wilhelm *et al.* Calendar aging of lithium-ion batteries, *Journal of the Electrochemical Society*, 163(9), A1872–A1880, 2016.

[25] N. I. Shchurov, S. I. Dedov, B. V. Malozyomov *et al.* Degradation of lithium-ion batteries in an electric transport complex, *Energies,* 14(23), 8072, 2021.

[26] J. Jaguemont, L. Boulon, and Y. Dubé, A comprehensive review of lithium-ion batteries used in hybrid and electric vehicles at cold temperatures, *Applied Energy,* 164, 99–114, 2016.

[27] B. E. Worku, S. Zheng, and B. Wang, Review of low-temperature lithium-ion battery progress: New battery system design imperative, *International Journal of Energy Research,* 46(11), 14609–14626, 2022.

[28] Y. Gao, T. Rojas, K. Wang *et al.* Low-temperature and high-rate-charging lithium metal batteries enabled by an electrochemically active monolayer-regulated interface, *Nature Energy,* 5(7), 534–542, 2020.

[29] W. Lin, M. Zhu, Y. Fan *et al.* Low temperature lithium-ion batteries electrolytes: Rational design, advancements, and future perspectives, *Journal of Alloys and Compounds,* 905, 164163, 2022.

[30] J. Sieg, J. Bandlow, T. Mitsch *et al.* Fast charging of an electric vehicle lithium-ion battery at the limit of the lithium deposition process, *Journal of Power Sources,* 427, 260–270, 2019.

[31] J. Zhang, H. Ge, Z. Li *et al.*, Internal heating of lithium-ion batteries using alternating current based on the heat generation model in frequency domain, *Journal of Power Sources,* 273, 1030–1037, 2015.

[32] X. Wu, W. Wang, Y. Sun *et al.*, Study on the capacity fading effect of low-rate charging on lithium-ion batteries in low-temperature environment, *World Electric Vehicle Journal,* 11(3), 55, 2020.

© 2025 World Scientific Publishing Company
https://doi.org/10.9789811282058_0008

## Chapter 8

# Power-to-X Technology for Decarbonizing Non-Land Transportation, Heating, and Chemical Industry

**Paul Rößner[*,‡] and Kai Peter Birke[†,§]**

*University of Stuttgart, Institute for Photovoltaics, Electrical Energy Storage Systems, Pfaffenwaldring, Stuttgart, Germany*

†*Fraunhofer IPA, Nobelstrasse 12, Stuttgart, Germany*

‡*paul.roessner@ipv.uni-stuttgart.de*

§*kai.peter.birke@ipa.fraunhofer.de*

This chapter provides an in-depth exploration of Power-to-X (PtX) technologies and their role in achieving a sustainable, fully electrified economy. It begins with an introduction to PtX, highlighting its various applications, advantages, and challenges. A comparative analysis with batteries clarifies the appropriate contexts for each technology, emphasizing complementary uses. The mobility sector is scrutinized, examining the efficiency and practicality of batteries versus e-fuels, with a focus on synergy effects and future policy implications. Efficiency chains for battery electric vehicles, hydrogen fuel cell vehicles, ammonia, and e-fuels are compared to underscore the importance of efficiency in climate goals. The role of PtX in decarbonizing the chemical industry is detailed, showcasing innovative production pathways and integration with renewable energy. The chapter examines plasma technology's advantages, efficiencies, and challenges within an electrified energy system. It also addresses the potential for sufficient renewable electricity, discussing energy capacities, efficiency, storage technologies, sector coupling, and international cooperation. In conclusion, the chapter presents a vision of an integrated energy system leveraging PtX technologies for a 100% renewable future, emphasizing their potential and the need for supportive policies.

## 8.1 An Introduction to Power-to-X Technologies

Maintaining the balance of supply and demand in an electricity grid requires constant adjustments to power generation and consumption. This balance becomes more challenging with the increased integration of renewable energy sources, which can vary in output. An energy system based on 100% renewable electricity requires short-term energy storage (5 h) of 0.18 TWh and long-term (17 days) energy storage of 26 TWh [1]. In 2020, the installed short-term energy storage capacity was around 1 GWh (0.001 TWh), and by 2024, it had grown to over 14 GWh (0.014 TWh), showing significant growth and dynamism in short-term energy storage capacity [2, 4]. This growth is primarily driven by lower production costs for batteries, leading to widespread application of the technology. Battery projects generate revenue with each charge and discharge cycle, making long-term storage less attractive due to the lower number of cycles and, consequently, lower revenue.

Batteries are an integrated system storing energy in the same device that produces the storable species. Storing energy for 17+ days requires a different approach. The production unit needs to be separated from the storage unit. This allows for separate scaling of power and capacity and enables applications and business cases that cannot be covered by batteries. The separation of production and storage can be achieved by processes that use renewable energy to produce molecules, which can be stored outside of the production unit. This can be viewed as a battery with a very high energy density and locally separated charge/discharge and storage units. This approach complements the currently growing field of battery applications and is known as Power-to-X (PtX).

The term "Power-to-X" is an evolution of the original term "Power-to-Gas" (PtG), coined by Michael Sterner [5, 6]. In addition to the conversion of power (i.e., renewable electricity) to gas (synthetic methane), the term now describes the conversion of electrical energy into various forms of energy or chemical compounds that can be stored, transported, and later reconverted into electricity or other energy services. This versatile concept plays a crucial role in the energy transition and the integration of renewable energy sources into the energy system. [7]

### 8.1.1 *Advantages of Power-to-X technologies*

The advantages of PtX technologies are multifaceted. PtX enables the optimal integration of renewable energy generation into the entire energy system, allowing for the effective utilization of these clean sources. Moreover, PtX technologies facilitate the storage and transport of energy over large distances and time periods, promoting the flexibility and efficiency of the energy system as a whole through sector coupling. By exclusively using renewable energy for the production of fuels and basic chemicals, PtX can significantly reduce greenhouse gas emissions, as fossil fuels and nuclear energy are not suitable for power generation in a clean energy system. Replacing fossil energy and feedstock sources is essential for a transformed climate-compatible energy and production system, and PtX provides the process technology to achieve this. Furthermore, the current fossil energy system experiences significant energy losses, with around two-thirds of primary energy being lost. An electrified system utilizing PtX technologies is much more efficient and requires less primary energy input compared to existing fossil-based system.

The key advantages of PtX technologies are as follows:

**Integration of renewable energies:** PtX enables the use of excess renewable energy when production exceeds current demand.

**Storage and transport:** PtX technologies allow for the storage and transport of energy over large distances and long periods.

**Sector coupling:** PtX promotes the interconnection of different energy sectors (electricity, heat, mobility, and industry), contributing to the flexibility and efficiency of the entire energy system.

**Reduction of greenhouse gas emissions:** By using renewable energies to produce fuels and chemical feedstocks, $CO_2$ emissions can be significantly reduced.

## 8.1.2 Challenges of Power-to-X technologies

The development and widespread adoption of PtX technologies face several key challenges. The technologies are still relatively expensive and require substantial further research and development to become economically competitive and viable at scale. Additionally, the conversion processes often incur significant energy losses, reducing the overall efficiency of the systems. This necessitates a profound change in decision-making, prioritizing direct electrification due to its superior efficiency. While water electrolysis includes one transformation step with losses in storage and transportation, it can still be a viable solution in the appropriate application scenario. If hydrogen is not suitable, it may be necessary to accept more transformation steps with the respective efficiency losses to produce e-fuels instead.

A significant challenge is the need for substantial investments in the necessary infrastructure for the large-scale production, transport, and utilization of PtX products. Overcoming these hurdles is crucial to enabling PtX to reach its full transformative potential and support the urgent transition to a more sustainable energy future in the face of the worsening climate crisis.

Policymakers need to establish a stable, predictable, and long-term policy framework to effectively direct investments from private and public actors toward PtX technologies. Given the urgent need to address the climate crisis, there is insufficient time and resources to rely solely on market forces and competition to drive the necessary transition. A comprehensive policy approach, including targeted incentives, regulations, and coordinated infrastructure development, is essential to accelerate the widespread adoption of these transformative technologies and support the transition to a sustainable energy future.

The key challenges of PtX technologies are as follows:

**Costs:** The technologies are still relatively expensive and require further research and development to become economically competitive. E-fuel production costs are estimated to decrease from 8 €/L to about 5.5 €/L, but they remain more costly than fossil alternatives [8].

**Efficiency:** The conversion processes often incur significant energy losses, reducing the overall energy efficiency. For instance, converting electricity to hydrogen and then to e-fuels can result in substantial energy loss, making the entire process less efficient compared to direct electrification. For example, water electrolysis can lose around 30% of the input energy during conversion.

**Infrastructure:** Developing the necessary infrastructure for the production of PtX products requires substantial investments. For example, establishing a network of hydrogen refueling stations for fuel cell vehicles demands significant capital outlay and coordination, which is currently lacking in many regions. This includes not only the refueling stations themselves but also the pipelines and storage facilities needed to support a hydrogen-based energy system.

Overcoming these hurdles is crucial to enabling PtX to reach its full transformative potential and support the urgent transition to a more sustainable energy future in the face of the worsening climate crisis.

## 8.2 Differentiating Power-to-X Technologies from Batteries

The capacity (how much energy can be stored) and power (how quickly energy can be delivered) of batteries scale differently. Increasing the capacity of a battery system generally involves adding more battery cells, which directly increases the energy stored but also adds weight and volume.

In PtX systems, capacity scaling can be more straightforward. For example, increasing hydrogen storage capacity might involve adding more storage tanks, a process that can be achieved with relatively simple infrastructure modifications. However, power scaling in PtX technologies can be more challenging, as it requires larger electrolyzes to handle higher power inputs. This scaling often demands significant capital investment and space. This leads to a significant advantage regarding their capability for long-term energy storage. Technologies such as PtG enable the storage of energy over extended periods, which is particularly useful for managing seasonal fluctuations in renewable energy production. Energy harvested during periods of high renewable generation can be stored and used later when production is low, ensuring a steady and reliable energy supply throughout the year.

Additionally, PtX products such as hydrogen and e-fuels exhibit versatile applications across various sectors, including industry, mobility, and heating. This versatility makes PtX a valuable tool for diversifying and stabilizing our energy portfolio. For example, hydrogen can be used directly as a fuel or converted into other forms such as synthetic methane or liquid fuels, providing flexibility in energy utilization. Furthermore, PtX technologies facilitate energy transport and infrastructure. Gases and liquid fuels can be transported relatively easily over long distances and integrated into existing infrastructures, such as natural gas networks. By enabling the integration of renewable energy into the energy system, PtX technologies help utilize excess renewable energy, ensuring that no energy is wasted and making it available during times of low production. This capability is vital for maximizing the efficiency and sustainability of the energy system.

Batteries, in contrast, offer a distinct and complementary set of advantages. One of the most significant benefits is their high efficiency. Batteries generally exhibit very high efficiencies, with minimal energy losses during storage and retrieval. This makes them highly effective for capturing and using energy with minimal waste. Another key advantage of batteries is their fast response times. Batteries can quickly store and release energy, making them ideal for short-term balancing measures and grid stabilization. This rapid response capability is crucial for maintaining grid stability, particularly as the share of intermittent renewable energy sources such as wind and solar increases.

Additionally, batteries benefit from technological maturity. Technologies like lithium-ion batteries are well-established and widely used, providing a reliable and proven option for energy storage. Their extensive use has also led to economies of scale, reducing costs and increasing accessibility.

Moreover, batteries offer modularity and scalability. Battery systems can be easily scaled and expanded modularly to meet various capacity requirements. This flexibility makes them suitable for a wide range of applications, from small-scale residential systems to large-scale grid storage solutions. However, the costs associated with large battery systems can be prohibitively high, making them less economically viable for extensive energy storage needs. Additionally, safety concerns, such as the risk of thermal runaway and fire hazards, pose challenges in deploying large-scale battery installations.

A clear example of these differences is seen in renewable energy integration. Batteries are excellent for short-term storage and rapid response, providing quick bursts of energy to stabilize the grid. In contrast, for long-term energy storage, such as storing excess solar energy generated during the summer for use in winter, PtX technologies like hydrogen storage are more suitable. Despite their lower round-trip efficiency, PtX technologies can store large amounts of energy over extended periods with minimal losses.

In conclusion, while batteries excel in high energy throughput and frequent cycling applications, PtX technologies offer advantages in scalability of capacity and long-term energy storage. Understanding these differences is crucial for optimizing their use in various scenarios within the energy landscape.

### 8.2.1 *Matching technology to scenario*

Choosing the appropriate technology depends heavily on the specific application scenario. Here are some scenarios and the best-suited technologies for each:

**Short-term grid stabilization and frequency regulation:** Batteries excel in short-term grid stabilization and frequency regulation due to their rapid response times and high efficiency. They can quickly address fluctuations in energy supply and demand, ensuring grid stability.

**Medium-term storage to bridge day-night cycles:** Batteries are ideal for medium-term storage, such as bridging day-night cycles. They efficiently store and release energy over several hours, making them well suited for managing daily variations in renewable energy production and consumption.

**Long-term storage to bridge seasonal fluctuations:** PtX technologies such as PtG are optimal for long-term energy storage to bridge seasonal fluctuations. These technologies can store energy for weeks or months, making them ideal for maintaining a steady energy supply during periods of low renewable energy production.

**Decarbonization of industry and heavy transport:** For the decarbonization of industry and heavy transport, PtX products such as hydrogen and e-fuels are the most suitable options. These products can be used in industrial processes and heavy transport applications, providing a viable alternative to fossil fuels and reducing greenhouse gas emissions.

**Local energy storage and use in households:** For local energy storage and use in households, batteries are the most practical solution. They can be easily installed in homes to store solar energy, enhancing self-consumption and potentially contributing to grid stability.

### 8.2.2 *Conclusion*

Ultimately, the choice of technology depends on the specific context and application. In an integrated energy system, PtX technologies and batteries can complement each other effectively, addressing various energy storage and utilization needs. A holistic approach that combines both technologies can enhance the efficiency and reliability of the entire energy system, advancing the energy transition. By leveraging the strengths of each technology, we can build a more resilient and sustainable energy future.

## 8.3 Decarbonizing the Mobility Sector

PtX products, particularly e-fuels and batteries, each offer distinct advantages for the mobility sector and can effectively complement each other to provide a sustainable and flexible mobility solution. This section explores various scenarios and considerations for integrating these technologies to enhance transportation. However, it is important to note that these technologies are not the sole solutions nor necessarily the best ones in every context. As highlighted earlier, enhancing overall system efficiency is critical to meeting climate goals. Achieving these targets requires more than merely transitioning to different fuels or powertrains. It demands a comprehensive overhaul of transportation habits and infrastructure. The most significant efficiency boost can be achieved through the expansion and enhancement of public transportation. Reducing the number of cars on the road in favor of expanded and improved public transport systems offers substantial potential for system-wide efficiency gains. Prioritizing public transport can significantly cut emissions, reduce energy consumption, and advance our climate objectives.

Batteries are well-suited for various applications within the mobility sector. Electric vehicles (EVs) excel in urban and regional transportation due to their high efficiency and zero local emissions, making them ideal for daily commutes and short trips within cities. Light commercial vehicles, such as those used for delivery services within urban areas, benefit from the efficiency and low operating costs of EVs, particularly given their short driving distances and frequent stops. Public transport systems stand to benefit significantly from battery technology. Electric buses and trams not only improve urban air quality but also offer a highly efficient solution for public transportation needs. Additionally, the rise of micromobility solutions, including e-bikes, e-scooters, and other small EVs, provides convenient and eco-friendly options for short distances and daily commutes.

**EVs:** Ideal for urban and regional traffic due to their high efficiency and zero local emissions.

**Light commercial vehicles:** Suitable for delivery traffic in urban areas due to short driving distances and frequent stops.

**Public transport:** Electric buses and trams are efficient and improve air quality in cities.

**Micromobility:** E-bikes, e-scooters, and other small electric vehicles are ideal for short distances and commutes.

Batteries offer several key advantages in the mobility sector. They are highly efficient, providing a direct conversion of electrical energy into driving power, which results in minimal energy losses. Operating costs are low, as the cost of energy for batteries is cheaper compared to traditional internal combustion engines, and maintenance requirements are significantly reduced. Moreover, electric vehicles produce no local emissions, which substantially reduces air pollution and noise levels in urban areas, contributing to a healthier environment.

**High energy efficiency:** Direct conversion of electrical energy into driving power with minimal energy losses.

**Low operating costs:** Lower energy costs and reduced maintenance compared to internal combustion engines.

**Zero local emissions:** Significant reduction in air pollution and noise levels.

However, batteries also face several limitations. One major challenge is their limited range compared to traditional internal combustion engines, which can be a significant concern for long-distance travel. Additionally, the widespread and reliable charging infrastructure required to support extensive use of electric vehicles is still under development in many areas. Furthermore, charging batteries takes longer than refueling with liquid fuels, which can be inconvenient for users who need quick turnaround times.

**Range limitation:** A limited range compared to traditional internal combustion engines, especially for long-distance travel.

**Charging infrastructure:** Requires extensive and reliable charging infrastructure to support widespread use.

**Charging times:** Longer charging times compared to refueling with liquid fuels.

E-fuels produced through PtX processes are particularly useful in specific applications. They are ideal for long-distance and heavy-duty transport, such as trucks, airplanes, and ships, where the weight and charging times of batteries pose significant challenges. Additionally, specialty vehicles, such as agricultural machinery and construction equipment, can benefit from e-fuels due to their specific requirements and operating conditions.

**Long-distance and heavy-duty transport:** Suitable for trucks, airplanes, and ships that travel long distances, where battery weight and charging times present significant challenges.

**High energy density:** Ideal for long-distance, non-land travel, and heavy loads due to the high energy density of liquid fuels.

**Long-term storage:** Suitable for seasonal energy storage and ensuring energy availability when needed.

However, e-fuels face several challenges. The production and use of e-fuels are less efficient compared to the direct use of electricity in batteries, leading to higher overall energy consumption. Currently, the costs of producing e-fuels are higher than those for fossil fuels and batteries, making them less economically attractive. Additionally, while e-fuels can be produced using renewable energy, their combustion still produces $CO_2$ emissions. Although these emissions can be neutralized if the production process uses captured $CO_2$, this requires meticulous management.

Furthermore, e-fuels are not yet commercially available on a large scale, with only a few demonstration plants existing worldwide. By 2035, around 60 new e-fuel projects have been announced, but only about 1% have secured final investment decisions. Collectively, these global projects would meet only about 10% of Germany's essential e-fuel needs for aviation, shipping, and the chemical industry. To accelerate the e-fuel market ramp-up, policymakers have the potential to implement mandatory quotas for e-fuels in aviation and shipping [9, 10].

**Lower efficiency:** Less efficient than the direct use of electricity in batteries, leading to higher energy consumption.

**Higher production costs:** Currently more expensive to produce than fossil fuels and batteries.

**$CO_2$ emissions:** Combustion produces $CO_2$ emissions, which can be neutralized if the production process utilizes $CO_2$.

## 8.4 Complementarity and Synergy Effects

To achieve a sustainable and efficient mobility sector, batteries and e-fuels can be used in complementary ways. In urban areas and for short distances, battery-powered vehicles are ideal due to the growing availability of charging infrastructure. For long-distance travel and heavy-duty vehicles, e-fuels offer a practical solution due to their high energy density and straightforward refueling capabilities. They also facilitate an immediate reduction in $CO_2$ emissions for existing internal combustion engine fleets, bridging the gap as electric mobility continues to expand.

Hybrid systems combining battery-electric drivetrains with internal combustion engines powered by e-fuels provide versatile solutions during the transition phase. Expanding charging infrastructure will support the broader adoption of electric vehicles, while utilizing excess renewable energy to produce e-fuels can serve as a long-term energy storage solution and meet peak demand requirements.

**Urban areas and short distances:** Battery-powered vehicles (EVs) are ideal for urban transportation and short commutes where charging infrastructure is readily available.

**Long-distance and heavy transport:** E-fuels are better suited for long-distance travel and heavy-duty vehicles where high energy density and easy refueling are crucial.

**Transitional solutions:** E-fuels enable an immediate reduction in $CO_2$ emissions in existing internal combustion engine fleets, bridging the gap as electric mobility continues to expand.

**Hybrid systems:** Combining battery-electric drivetrains with internal combustion engines powered by e-fuels provides versatile solutions during the transition phase.

## 8.5 Future Outlook and Policy Considerations

The future of mobility is likely to involve a combination of battery-electric vehicles and e-fuels. Policymakers and industry stakeholders must collaborate to establish a supportive framework that fosters investment and innovation in both technologies. Financial incentives are crucial for accelerating adoption and enhancing the competitiveness of both electric vehicles and e-fuels. Ongoing investment in research and development is also essential for improving the efficiency and reducing the costs of both batteries and e-fuels. Clear and consistent regulatory frameworks are crucial for providing certainty to investors and driving the transition to a more sustainable mobility sector. Additionally, educating the public about the benefits and capabilities of these technologies can enhance acceptance and encourage broader adoption.

PtX technologies can be viewed as a different kind of "battery" that stores energy in various forms, which can then be utilized across multiple applications. PtX involves converting surplus renewable electricity into alternative energy carriers, such as hydrogen, synthetic methane, or e-fuels. These energy carriers can be stored for extended periods and used across different sectors, including transportation, heating, and industrial processes.

The versatility of PtX products is a significant advantage. Once energy is stored in a chemical form, it can be utilized across a wide range of applications, each with specific energy requirements. For instance, synthetic methane can be used in solid oxide fuel cells for power generation, injected into the natural gas grid, or applied in various industrial processes. This flexibility enables PtX technologies to address energy requirements in sectors where direct electrification may be challenging or inefficient.

Overall, while batteries are ideal for applications requiring frequent cycling and high energy density, PtX technologies provide a complementary solution by facilitating long-term storage and versatile utilization of renewable energy across various sectors. This combined approach can help achieve a more resilient and sustainable energy system, efficiently leveraging the strengths of both technologies to meet the diverse energy demands of the future.

The integration of batteries and e-fuels in the mobility sector offers a flexible, efficient, and sustainable solution for transportation needs. While batteries are ideal for urban areas and short to medium distances, e-fuels provide a practical option for long-distance and heavy-duty transport. The choice of technology depends heavily on the specific use case, and a holistic strategy that integrates both technologies can effectively support the goals of energy transition and emission reduction. By leveraging the strengths of each technology, we can create a more resilient and sustainable mobility future.

## 8.6 Comparative Analysis of Solutions for the Transportation Sector: Overall Efficiency

Comparing the overall efficiencies of various technologies for the transportation sector – **batteries, hydrogen, ammonia, and e-fuels** – can be complex due to the various conversion stages and associated energy losses involved. This section examines the efficiencies of each technology, tracing the process from the generation of primary electricity to its use in the vehicle.

### 8.6.1 *Motivation for efficiency in achieving climate goals*

As emphasized earlier, improving the overall efficiency of our energy systems is critical for meeting climate targets. Simply switching fuels or powertrains is insufficient. The most substantial efficiency gains can be achieved by prioritizing public transportation and reducing the number of cars on the road. Public transportation systems can substantially cut emissions, reduce energy consumption, and make a significant contribution to our climate objectives. However, for the remaining vehicular traffic, a comparative understanding of different energy carriers and technologies is essential for optimizing the sector's efficiency.

### 8.6.2 *Efficiency chain for battery electric vehicles*

Battery electric vehicles (BEVs) are highly efficient in converting renewable electricity into mechanical energy for transportation. The efficiency chain for BEVs encompasses several stages.

The process begins with the generation of renewable electricity, which starts at 100% efficiency. This electricity is then transmitted and distributed, with an approximate loss of 5%, resulting in an efficiency of about 95%. Charging the battery incurs further losses, typically around 10%, leading to an efficiency of 90% during this stage. The batteries themselves experience storage losses, averaging 5%, yielding an effective storage efficiency of 95%. Finally, the discharge process and the efficiency of the electric motor are around 90%. When combined, these stages result in an overall efficiency of approximately 73%:

- generation of renewable electricity: 100%,
- transmission and distribution: ~95%,
- battery charging: ~90%,
- storage losses in the battery: ~95%,
- discharge and motor efficiency: ~90%,
- *overall efficiency: ~73%.*

### 8.6.3 *Efficiency chain for hydrogen fuel cell vehicles*

Hydrogen fuel cell electric vehicles (FCEVs) involve several stages to convert renewable electricity into hydrogen. First, renewable electricity is used to produce hydrogen through electrolysis. This hydrogen is then compressed or liquefied for storage and transport. Finally, the hydrogen is utilized in a fuel cell to generate electricity for propulsion.

For hydrogen FCEVs, the efficiency chain begins with the generation of renewable electricity, which is 100%. Electrolysis, the process that converts electricity into hydrogen, has an efficiency of about 70%. Compressing, liquefying, and storing the hydrogen achieves an efficiency of about 90%.

Transport and refueling contribute an efficiency of roughly 95%. Finally, the fuel cell's efficiency in converting hydrogen back into electricity is around 50%. Combining these stages results in an overall efficiency of about 30%:

- generation of renewable electricity: 100%,
- electrolysis (hydrogen production): ~70%,
- compression/liquefaction and storage: ~90%,
- transport and refueling: ~95%,
- fuel cell efficiency: ~50%,
- *overall efficiency: ~30%.*

### 8.6.4 *Efficiency chain of ammonia as an energy carrier*

Using ammonia as an energy carrier involves several stages: converting renewable electricity into hydrogen through electrolysis, synthesizing ammonia, and then using it in either ammonia fuel cells or combustion engines.

The efficiency chain for ammonia starts with the generation of renewable electricity at 100%. Electrolysis for hydrogen production operates at about 70% efficiency. The synthesis of ammonia using the Haber–Bosch process achieves about 70% efficiency. Transport and storage of ammonia are highly efficient, around 95%. Finally, when used in fuel cells or combustion engines, ammonia's efficiency is about 40%. Combining these stages results in an overall efficiency of approximately 19%.

- generation of renewable electricity: 100%,
- electrolysis (hydrogen production): ~70%,
- ammonia synthesis (Haber–Bosch process): ~70%,
- transport and storage: ~95%,
- use in ammonia fuel cells or combustion engines: ~40%,
- *overall efficiency: ~19%.*

### 8.6.5 *Efficiency chain for e-fuels*

E-fuels involve the following stages: converting renewable electricity into hydrogen through electrolysis, synthesizing e-fuels using captured $CO_2$, and combusting these fuels in internal combustion engines.

236 *P. Rößner & K. P. Birke*

For e-fuels, the efficiency chain begins with the generation of renewable electricity at 100%. Electrolysis for hydrogen production operates at about 70% efficiency. The synthesis of e-fuels from hydrogen and $CO_2$ achieves around 60% efficiency. Transport and storage efficiencies are roughly 95% efficient. However, the combustion of e-fuels in engines is about 25% efficient. This results in an overall efficiency of approximately 10%.

- generation of renewable electricity: 100%,
- electrolysis (hydrogen production): ~70%,
- e-fuels production: ~60%,
- transport and storage: ~95%,
- combustion in engines: ~25%,
- *overall efficiency: ~10%.*

### 8.6.6 *Summary of overall efficiencies and conclusion*

In summary, the overall efficiencies of the different technologies are as follows:

- BEVs: ~73%,
- FCEVs: ~30%,
- ammonia: ~19%,
- e-fuels: ~10%.

Direct electric drives (batteries) exhibit the highest overall efficiency, followed by hydrogen, ammonia, and e-fuels. This indicates that BEVs enable the most efficient use of renewable electricity. Hydrogen, ammonia, and e-fuels, although less efficient as energy carriers and storage solutions, offer other advantages such as higher energy density and easier transport over long distances.

The choice of technology depends on the specific requirements of the application. For urban transport and short distances, BEVs are the most efficient. Hydrogen is ideal where rapid refueling and long ranges are essential, such as in heavy-duty transport. Ammonia and e-fuels are suitable for long-haul flights and maritime transport, where high energy densities and existing infrastructure are beneficial.

Figure 8.1 illustrates the overall efficiencies of different energy technologies used in transportation and their corresponding energy conversion

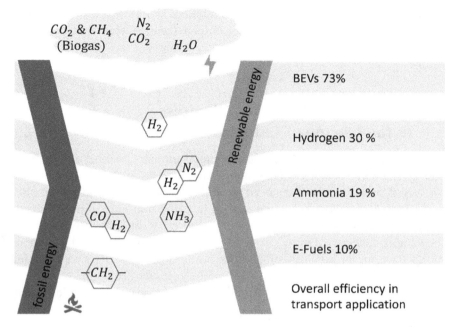

**Figure 8.1.** Efficiency comparison of renewable energy technologies for transportation.

processes. BEVs stand out with the highest efficiency at approximately 73%. This high efficiency is attributed to the direct use of electricity for propulsion, which minimizes energy losses. In contrast, FCEVs have a significantly lower efficiency of around 30%. The energy conversion process for FCEVs involves using renewable electricity to produce hydrogen through electrolysis, followed by generating electricity in the fuel cell, resulting in considerable energy loss at each stage.

Ammonia, as an energy carrier, demonstrates an efficiency of about 19%. The process involves converting renewable electricity into hydrogen and then synthesizing ammonia, adding further energy losses. E-fuels exhibit the lowest efficiency, approximately 10%. The production of e-fuels involves several energy-intensive steps, including electrolysis to produce hydrogen and subsequent chemical synthesis processes to create liquid fuels from $CO_2$ and hydrogen. These multiple stages contribute to significant energy losses, resulting in a notably lower overall efficiency.

Figure 8.1 highlights the progressive decline in efficiency with each downstream energy conversion step. Beginning with the direct use of renewable electricity in BEVs, efficiency decreases as the process involves more complex chemical conversions in FCEVs, ammonia, and e-fuels. The cumulative energy losses emphasize the importance of prioritizing the most efficient technologies to achieve a sustainable energy future.

## 8.7 The Role of Power-to-X Processes in the Chemical Industry

The chemical industry, alongside the metal industry, is one of the most energy-intensive sectors in Germany. A distinguishing feature of this sector is that energy carriers are used not only for generating energy but also as feedstock for chemical processes. In fact, the non-energy use of these carriers surpasses their use as an energy source, with 86% of non-energy feedstocks currently derived from fossil raw materials, particularly those based on mineral oil. Conversely, the sector's process energy needs are primarily met with natural gas (42%) and electricity (26%) [11].

PtX processes offer innovative pathways to produce chemicals and fuels from renewable energy sources, thereby reducing reliance on fossil resources and lowering $CO_2$ emissions. This section explores the role of PtX in decarbonizing the chemical industry, enhancing energy efficiency, integrating renewable energy, and developing new production methods. It also addresses the challenges associated with PtX and potential solutions.

### 8.7.1 *Decarbonization of production*

The chemical industry's substantial $CO_2$ emissions and reliance on fossil fuels make it an ideal candidate for decarbonization through PtX processes. These technologies facilitate the production of chemical feedstocks and fuels from renewable energy, significantly reducing carbon footprints. By capturing and utilizing $CO_2$ (CCU), we can convert it into valuable raw materials for e-fuels and chemicals, effectively closing the carbon loop. Additionally, green hydrogen, produced through electrolysis using renewable electricity, serves as a critical feedstock for various chemical processes, including the production of ammonia and methanol. This approach not only minimizes $CO_2$ emissions but also encourages the use of sustainable raw materials [12].

**CCU:** $CO_2$ can be used as a feedstock for producing e-fuels and chemicals, closing the carbon cycle.

**Production of hydrogen via electrolysis:** Green hydrogen, produced from renewable electricity, can be used as a feedstock for various chemical processes, including the production of ammonia and methanol.

### 8.7.2 *Enhancing energy efficiency*

High-temperature plasma processes, for example, can be employed to split molecules like $CO_2$ and $CH_4$, creating more efficient pathways for chemical production. Furthermore, integrating renewable electricity allows PtX plants to store excess energy, which can be utilized during periods of high demand, thereby boosting the overall energy efficiency of the system.

**Plasma processes:** Plasma technologies, such as microwave and gliding arc plasmas, can effectively split molecules like $CO_2$ and $CH_4$, facilitating more efficient production pathways for chemicals.

**Integration of renewable electricity:** By harnessing excess renewable electricity, PtX plants can store energy and deploy it during high-demand periods, enhancing the overall energy efficiency of the system.

### 8.7.3 *Flexibility and integration of renewable energy*

PtX technologies provide diverse opportunities for integrating renewable energy sources into chemical production, especially considering the intermittent nature of wind and solar power.

PtX plants can serve as flexible consumers, operating during periods of excess renewable energy and shutting down during shortages, thereby enhancing grid stability. Additionally, by converting electricity into chemical energy carriers like hydrogen or e-fuels, PtX processes serve as energy storage systems, efficiently managing seasonal fluctuations in renewable energy supply.

**Load management:** PtX plants can operate as flexible consumers, activating during renewable energy surpluses and scaling back during shortages, thereby supporting grid stability.

**Energy storage:** By converting electricity into chemical energy carriers like hydrogen or e-fuels, PtX processes serve as energy storage systems, efficiently managing seasonal fluctuations.

### 8.7.4 *Innovative production pathways*

PtX technologies create new, sustainable production pathways for chemicals that have traditionally been derived from fossil resources.

E-fuels provide climate-neutral alternatives to conventional fuels, particularly in sectors where electrification is challenging, such as aviation and maritime transport. Additionally, green ammonia and methanol, produced using renewable hydrogen and $CO_2$ or nitrogen, offer versatile applications in the chemical industry, presenting sustainable options for essential chemical feedstocks.

**E-fuels:** Methanol, ammonia, or synthetic diesel can serve as climate-neutral alternatives to conventional fuels, particularly in sectors where electrification is challenging, such as aviation and maritime transport.

**Green ammonia and methanol:** These basic chemicals, produced using renewable hydrogen and $CO_2$ or nitrogen, have diverse applications in the chemical industry. They are used in fertilizer production, the creation of fine chemicals for the cosmetic industry, and the production of monomers for plastics.

### 8.7.5 *Challenges and solutions*

While PtX technologies hold great promise, they face several challenges, including high costs, technological maturity, and infrastructure requirements.

**Challenges:**

- *Costs*: The initial investments and operating costs for PtX plants are currently high due to the complexity of the required hardware and the cost of electricity, which remains a significant expense.
- *Technological maturity*: Many PtX technologies are still in the development or pilot phase. They face competition from established processes that rely on relatively cheap fossil fuels, with limited investment in PtX advancements in recent years.

- *Infrastructure*: Existing infrastructure must be adapted or expanded to accommodate PtX technologies. For example, refineries and integrated chemical sites need to transition from processing fossil feedstocks to renewable feedstocks and electrified processes.

**Solutions:**

- *Research and development*: Investing in R&D is essential to improving the efficiency and cost-effectiveness of PtX technologies. The energy transition requires investments of €1,200 billion over the next decade, equating to €120 billion annually. This amounts to roughly 0.3% of Germany's annual gross domestic product each year for 10 years [13].
- *Policy support*: Government incentives and supportive policies, such as carbon pricing, can facilitate the market introduction and scaling of PtX technologies. Reliable, long-term commitments are crucial to provide certainty to investors, companies, and end users, ensuring that the path to a green transition remains consistent.
- *Cooperation and partnerships*: Collaboration between industry, research institutions, and governments is essential for accelerating the development and implementation of PtX technologies.

### 8.7.6 *Conclusion*

PtX technologies are pivotal for transitioning the chemical industry toward more sustainable and climate-friendly production methods. They present diverse opportunities to efficiently utilize renewable energy, reduce $CO_2$ emissions, and decrease reliance on fossil resources. Addressing the technical and economic challenges associated with PtX processes will be essential for their widespread adoption and significant contribution to global climate goals.

## 8.8 The Role of Plasma Technology in an Electrified Energy System

The electrification of the (petro)chemical industry represents one of the greatest challenges of the 21st century. Alongside the metal industry, the base chemicals sector is one of the most energy-intensive industries in Germany. Plasma technology is gaining increasing interest as a versatile solution for integrating with fluctuating renewable electricity. This section explores the role of plasma technologies within an electrified energy system, with a focus on their applications in both the transportation sector and industrial processes [14, 15].

### 8.8.1 *Advantages of plasma technology*

Plasma technology offers numerous advantages, positioning it as a promising solution for decarbonizing the chemical industry. One key benefit is its adaptability to fluctuating energy sources. Plasma processes can be rapidly started and stopped, making them well suited for integration with renewable energy sources, such as solar and wind power. Furthermore, plasma technology is versatile in its chemical conversions. It can support a wide range of reactions, including the splitting of $CO_2$, the conversion of $CH_4$, and the fixation of $N_2$. This flexibility is crucial for reducing carbon emissions in the chemical industry. Additionally, plasma processes generate highly reactive species, such as radicals and ions, which can enhance the efficiency of chemical reactions, particularly those that would otherwise require high temperatures and pressures.

### 8.8.2 *Comparison of overall efficiencies and costs*

When comparing the efficiencies and costs of plasma technology across different applications, several key points emerge. Plasma technology can effectively split $CO_2$ into CO and $O_2$, with CO then used in processes such as the Fischer–Tropsch synthesis to produce liquid fuels. Under optimal conditions, the overall efficiency of this process is approximately 60–70%. For methane conversion, plasma can efficiently transform methane into hydrogen and ethylene, potentially replacing conventional methods like steam reforming, with an efficiency of about 50–60%. In the case of nitrogen fixation, plasma technology can directly convert nitrogen into ammonia or NO$x$, providing an alternative to energy-intensive processes such as the Haber–Bosch process, with an efficiency of around 40–50%.

### 8.8.3 *Challenges and improvements*

Despite its advantages, plasma technology faces several challenges that must be addressed through ongoing research and innovation. Improving both energy efficiency and conversion efficiency is critical. This can be achieved by optimizing plasma reactors, enhancing gas flow dynamics, and implementing rapid quenching of products. Innovative approaches, such as pulsed plasmas or advanced nozzle configurations for rapid plasma cooling, could further boost efficiency. Another challenge is product selectivity. Combining plasma with catalysts can increase the selectivity of produced chemicals, though this remains complex and requires further study. Plasma-specific catalysis, tailored to the unique environment of plasmas, could offer novel catalytic effects.

### 8.8.4 *Applications in the transportation sector*

Plasma technology also holds significant potential for the transportation sector. By utilizing plasma technology to convert $CO_2$ and methane, e-fuels and chemical feedstocks can be produced. These can be used in conventional combustion engines or in fuel cell vehicles. Moreover, plasma-based methods for methane splitting could provide a carbon-neutral hydrogen source, suitable for use in fuel cell vehicles. This presents a sustainable option for the transportation sector, contributing to the reduction of greenhouse gas emissions.

### 8.8.5 *Conclusion*

Plasma technology presents promising opportunities for integration into an electrified energy system. It is particularly valuable for the chemical industry and the transportation sector, offering flexibility to respond to renewable energy sources and supporting various chemical conversion processes. Despite existing challenges related to efficiency and product selectivity, plasma processes have the potential to replace traditional, $CO_2$-intensive methods, thus contributing significantly to the decarbonization of the industry. Continuous research and development in this area will be crucial to fully exploit the benefits of plasma technology.

The question of whether there will be enough renewable electricity to support a fully electric economy is complex and depends on many factors, including technological advancements, political measures, infrastructure investments, and the rate at which renewable energy is expanded. Key considerations include improvements in the efficiency and performance of renewable energy technologies, advancements in energy storage and grid management, supportive government policies, and substantial investments in infrastructure to expand the capacity of renewable energy.

## 8.9 Evaluating the Future Availability of Renewable Electricity

The question of whether there will be sufficient renewable electricity to support a fully electrified economy is complex and multifaceted. Achieving this goal depends on numerous factors, including the expansion of renewable energy capacities, improvements in energy efficiency, development of storage technologies, sector coupling, international cooperation, and supportive political frameworks. However, various projections indicate that electricity demand may range from 1,200 to 1,500 TWh [7, 16] by 2045. The potential production capacity for renewable energy is estimated at approximately 1,200 to 1,400 TWh [7, 17].

Several technologies can increase the renewable energy share in the German energy system. A feasible transition strategy for achieving 100% renewable energy across all energy sectors has been outlined, emphasizing the importance of energy savings to remain within sustainable resource limits for renewable electricity and biomass. The heating sector, in particular, presents significant energy-saving potential, with additional savings achievable in the industrial and electricity sectors. Technologies such as electric vehicles, heat pumps, and electrolyzers not only improve energy system efficiency but also enhance system flexibility, facilitating the integration of additional renewable electricity. Consequently, it is imperative to deploy all viable renewable electricity resources to meet the substantial electrification needs of all energy sectors, and the data indicate that this is feasible.

### 8.9.1 *Expanding renewable energy capacities*

Transitioning the entire economy to electricity requires a substantial increase in renewable energy capacity across various sources. Solar energy, for instance, requires extensive deployment of photovoltaic systems on rooftops, in urban areas, and in large solar parks. Similarly, expanding wind energy, both onshore and offshore, is crucial. Hydropower can be harnessed effectively through rivers, dams, and ocean currents. Bioenergy, derived from the sustainable use of biomass, can serve as a valuable supplementary energy source. Additionally, geothermal energy presents significant potential by tapping into the Earth's thermal energy for both electricity generation and heating.

The extensive deployment of these renewable energy sources involves not only the installation but also the development of a resilient and scalable infrastructure. Investments in new technologies and the upgrading of existing grids are necessary to manage the increased load and ensure stable supply. Furthermore, policy measures to incentivize the adoption and integration of renewable energy systems are crucial for accelerating this transition.

To transition the entire economy to electricity, a substantial increase in renewable energy capacity is essential. This involves a diverse array of energy sources:

**Solar energy:** Extensive deployment of photovoltaic systems on rooftops, within urban areas, and in large solar parks is needed to increase capacity from 90 GW in 2024 to 320 GW in 2045 [4].

**Wind energy:** Expansion of wind farms is crucial, with onshore capacity increasing from 64 GW in 2024 to 199 GW in 2045, and offshore capacity growing from 7.5 GW in 2024 to 66 GW in 2045.

**Hydropower:** The utilization of rivers, dams, and ocean currents is nearly at its maximum potential of 100%, with no significant increase expected [2].

### 8.9.2 *Energy efficiency and savings*

Improving energy efficiency across all sectors can significantly reduce the overall energy demand, making the goal of a fully electrified economy more achievable. According to a Fraunhofer study [2], societal acceptance of new technologies and changes in behavior play a significant role. For example, adapting to the new routine of charging an electric vehicle, as opposed to the traditional practice of refueling at a gas station, is an important factor. The German government has set specific targets for achieving a renewable energy system and net-zero emissions by 2045. These goals include reaching an 80% share of renewable electricity production, 50% climate-neutral heating, 6 million installed heat pumps, 15 million BEV, and 10 GW of electrolyzer capacity. In the building sector, measures such as improved insulation, energy-efficient heating systems, and smart control systems can drastically cut energy consumption. In the industrial sector, benefits can be gained through more efficient production processes, waste heat recovery, and optimized energy use. In transportation, adopting more efficient vehicles, promoting public transport, and implementing sharing concepts can collectively contribute to substantial energy savings.

Efforts to enhance energy efficiency must also include comprehensive regulatory frameworks that set high efficiency standards for appliances, buildings, and vehicles. Public awareness campaigns can further encourage individuals and businesses to adopt energy-saving practices, thereby reducing the overall energy footprint.

Increasing energy efficiency across all sectors can significantly reduce overall energy demand:

**Building sector:** Improved insulation, energy-efficient heating systems, and smart control systems, such as smart meters and bidirectional charging of BEVs.

**Industry:** More efficient production processes, waste heat recovery, and optimized energy use.

**Transportation:** Promotion of more efficient transport, including public transport and sharing concepts.

### 8.9.3 *Flexibility and storage technologies*

Given the intermittent nature of renewable energy sources, developing flexibility and storage solutions is crucial. Battery storage systems are essential for providing short- to medium-term storage, stabilizing the grid, and covering peak demands. Pumped-storage power plants, a proven technology for large-scale energy storage, also play a significant role in managing energy supply and demand. Additionally, PtX technologies, which convert excess electricity into hydrogen, e-fuels, or heat, offer a versatile solution for long-term storage. Smart grids are another critical component, designed to manage and distribute electricity efficiently. They help balance supply and demand in real time, further enhancing the integration of renewable energies.

The development and deployment of these storage technologies and flexible systems will require significant investment and innovation. For example, improvements in battery technology can boost storage capacity and efficiency, while smart grid technology can enhance the resilience and reliability of electricity networks.

Given the intermittent nature of renewable energy sources, flexibility and storage solutions are crucial:

**Battery storage:** Short- to medium-term storage (<5 h) to stabilize the grid and cover peak demands.

**Pumped-storage power plants:** Proven technology for large-scale energy storage.

**PtX:** Conversion of excess electricity into hydrogen, e-fuels, or heat.

**Smart grids:** Intelligent electricity networks to better integrate renewable energies and manage consumption. Demand side control is a powerful way to minimize redundancy and achieve an efficient overall system. However, the necessary communication between producing units and consuming units is complex, as public backlash was observed after announcing flexible electricity market designs by the government in 2023 [18].

### 8.9.4 *Sector coupling*

Integrating different sectors – electricity, heat, transportation, and industry – can lead to more efficient use of renewable energy. The electrification of transportation, for instance, involves using electricity from renewable sources to power electric vehicles, thereby reducing reliance on fossil fuels. In the industrial sector, shifting from fossil-fuel-driven processes to electric ones can further decarbonize production. Power-to-Heat technologies, which use excess electricity to generate heat for residential and industrial applications, can help balance the grid and ensure energy is used efficiently.

Sector coupling not only improves energy efficiency but also provides greater flexibility in how and when energy is used. This interconnected approach can help mitigate the variability of renewable energy sources and ensure a more stable and sustainable energy system.

Integrating different sectors – electricity, heat, transportation, and industry – allows for more efficient use of renewable energy.

### 8.9.5 *International cooperation and grid expansion*

Enhanced international cooperation and network expansion can significantly increase energy security and efficiency. Within Europe, better interconnected grids can balance production and demand across the continent, making it easier to integrate renewable energy. Long-distance electricity transmission lines can connect regions with high renewable potential to major consumption centers, ensuring a steady supply of green energy.

International collaboration can also foster the exchange of best practices and technological innovations, accelerating the transition to renewable energy globally. Joint ventures and partnerships can leverage resources and expertise to develop and deploy renewable energy projects more effectively.

Enhanced international cooperation and network expansion can increase energy security:

**European energy union:** Better interconnected grids within Europe to balance production and demand.

**International power lines:** Long-distance electricity transmission lines connecting regions with high renewable potential to consumption centers.

### 8.9.6 *Forecasts and scenarios*

Various forecasts and scenarios suggest that operating a fully electrified economy based on renewable energy is theoretically feasible, provided that certain conditions are met. Continuous expansion of renewable energy capacity is crucial, which demands ambitious targets and consistent implementation. Innovations in storage technology, grid management, and energy efficiency will be pivotal in supporting this shift. Political backing, including clear frameworks, subsidies, and support programs, is essential to drive the energy transition forward. Moreover, societal acceptance and support from both the public and businesses are vital to ensure a smooth and successful transition.

Forecasts and scenarios indicate operating a fully electrified economy based on renewable energy is theoretically possible if certain conditions are met:

**Continuous expansion of renewable energy:** Ambitious expansion targets must be consistently pursued.

**Innovations and technologies:** Advances in storage technology, grid management, and energy efficiency are crucial.

**Political backing:** Clear political frameworks, subsidies, and support programs are essential for the energy transition.

**Societal acceptance:** Support and acceptance from the public and businesses are vital for a smooth transition.

### 8.9.7 *Conclusion*

There is potential for sufficient renewable electricity to support a fully electrified economy. However, this requires significant efforts in expanding renewable energy capacities, improving energy efficiency, implementing storage solutions, and comprehensive sector coupling. Successful implementation heavily relies on political support, technological advancements, and societal engagement. With coordinated efforts across all these areas, it is possible to achieve a sustainable, electrified future.

## 8.10 Concluding Insights on Power-to-X Technologies

In a future where renewable electricity forms the foundation of our energy system, the importance of PtX technologies and plasma processes is paramount. These innovations represent not only a departure from fossil fuels but also a transition to a more efficient, sustainable, and resilient energy landscape.

**Efficiency comparisons:** Plasma technology demonstrates promising efficiency rates across various applications. For instance, plasma processes can achieve a $CO_2$ splitting efficiency of about 60–70% under optimal conditions, converting $CO_2$ into CO and $O_2$, with CO serving as a valuable feedstock for e-fuel production through processes like Fischer–Tropsch synthesis. Similarly, methane conversion via plasma can yield an efficiency of approximately 50–60%, transforming methane into hydrogen and ethylene, thus providing an alternative to traditional steam reforming. Additionally, plasma can achieve efficiencies of about 40–50% in nitrogen fixation, directly producing ammonia or NO$x$, potentially replacing the energy-intensive Haber–Bosch process.

These efficiencies, although still evolving, highlight the potential for plasma technology to significantly decarbonize the chemical sector. Integrating these processes with renewable energy sources can further enhance the overall efficiency and sustainability of our energy systems.

### 8.10.1 *An integrated energy system for 100% renewable supply*

A fully renewable energy supply requires not only the expansion of renewable energy capacities but also a comprehensive approach that includes energy efficiency improvements, effective storage solutions, and sector coupling.

**Expanding renewable capacities:** Significant scaling up is required for solar, wind, hydropower, bioenergy, and geothermal energy. Solar energy should be deployed extensively in both urban and rural areas, while wind energy expansion should include both onshore and offshore installations. Hydropower, bioenergy, and geothermal sources will complement these efforts, providing a diverse and resilient renewable energy mix.

**Improving energy efficiency:** Maximizing energy efficiency across all sectors – buildings, industry, and transportation – is crucial. This involves adopting advanced insulation techniques, energy-efficient appliances, optimized industrial processes, and efficient transportation systems for both public and private use.

**Developing storage solutions:** Effective storage solutions are critical due to the intermittent nature of renewable energy. Batteries, pumped-storage power plants, and PtX technologies can store excess energy and release it when needed, ensuring a stable and reliable energy supply. This capability is particularly important during periods of extended low renewable generation, such as in winter with limited wind and solar resources. Smart grids will enhance the integration of renewable sources by facilitating real-time management of energy flows.

**Sector coupling:** Integrating the electricity, heat, transportation, and industrial sectors will enhance the efficient use of renewable energy.

Electrifying transportation and industry, along with implementing Power-to-Heat technologies, will ensure that renewable energy is effectively utilized across all areas of the economy.

**International cooperation and grid expansion:** Strengthened international cooperation and expanded electricity grids will boost energy security and efficiency. European energy networks can balance production and consumption across borders, while long-distance transmission lines can connect regions with high renewable potential to major consumption centers.

### 8.10.2 *The vision of a renewable energy future*

Imagine a world where factories operate on clean, renewable electricity and chemical feedstocks are sourced from water, air, and $CO_2$ instead of fossil fuels. This vision is achievable through PtX technologies, which can convert renewable energy into chemical products and fuels. Such a transformation has the potential to revolutionize an industry traditionally seen as challenging to decarbonize. The chemical industry, a cornerstone of the modern economy, could shift from being a major $CO_2$ emitter to a pivotal player in combating climate change [19].

**PtX technologies and their potential:** PtX processes enable us to reframe $CO_2$ as a valuable resource rather than a waste product. Green hydrogen, produced through the electrolysis of water using renewable electricity, serves both as a clean energy carrier and a feedstock for various chemical processes. PtX technologies can generate e-fuels and chemicals that replace fossil fuels in sectors challenging to electrify, such as aviation and maritime transport.

**Overcoming challenges:** Despite the current high costs and the developmental stage of many PtX processes, historical trends suggest that technological advancements can surpass expectations. With strategic investments in research and development, robust political support, and international cooperation, PtX technologies can advance to maturity and achieve large-scale deployment.

**A connected energy system:** Integrating PtX into our energy infrastructure can fundamentally change our approach to energy management. PtX plants can serve as flexible consumers of renewable energy, enhancing grid stability. During periods of excess electricity production, these plants can convert surplus energy into chemical carriers for storage. This not only boosts system efficiency but also reduces dependence on fossil energy storage.

**Envisioning a sustainable chemical industry:** PtX processes represent more than just technological innovations. They offer a pathway to a sustainable, climate-friendly future. By converting renewable energy into chemical products and fuels, PtX technologies have the potential to revolutionize an industry traditionally seen as challenging to decarbonize. The chemical industry, a cornerstone of the modern economy, could shift from being a major $CO_2$ emitter to a pivotal player in combating climate change.

### 8.10.3 *The potential of PtX technologies*

Looking ahead, the future of the chemical industry should be characterized by electric power, sustainability, and innovation. PtX technologies offer a transformative opportunity to align our economy with environmental goals. Imagine a world where the chemical industry is not merely a contributor to the problem but an integral part of the solution – a world where clean, renewable energy drives our production and consumption processes. This vision is within our reach, and each advancement in PtX technologies brings us closer to realizing it.

In conclusion, there is considerable potential for renewable electricity to support a fully electrified economy. Realizing this potential requires concerted efforts to expand renewable energy capacities, improve energy efficiency, develop advanced storage solutions, and implement comprehensive sector coupling.

Achieving this vision hinges on robust political support, ongoing technological innovation, and active societal engagement. By addressing these challenges directly, we can create a sustainable, electrified future that meets the energy demands of a modern economy while minimizing environmental impact.

## References

[1]  VDI ETG, Energiespeicher für die Energiewende, 2012.
[2]  Fraunhofer ISE, Wege zu einem klimaneutralen Energiesystem, 2021.
[3]  H. W. J. K. H. Brandes, Wege zu einem klimaneutralen Energiesystem, Fraunhofer ISE, 2021.

[4]   Fraunhofer-ISE, Energy-Charts, Available: https://www.energy-charts. info/charts/installed_power/chart.htm?l=de&c=DE, 2024.

[5]   S. Sterner, Power-to-gas and power-to-x – the history and results of developing a new storage concept, *MDPI Energies*, 2021.

[6]   N. S. Estermann, Power-to-gas systems for absorbing excess solar power in electricity distribution networks, *International Journal of Hydrogen Energy*, 41(32), 13950–13959, 2016.

[7]   F. Ausfelder and D. D. Tran, Options for a sustainable energy system with power-to-X technologies, federal ministry of education and research, 2022.

[8]   Kaufmann, Rößner, Renninger, Lambarth, Raab, Stein, Seithümmer, Birke, Techno-economic potential of plasma-based $CO_2$ splitting in power-to-liquid plants, *MDPI Applied Sciences*, 13, 4839, 2023.

[9]   O. Ueckerdt, E-fuels – Aktueller stand und projektionen, Potsdam-Institut für Klimafolgenforschung (PIK), 2023.

[10]  Ueckerdt, Bauer, Dirnaichner, Everall, Sacchi, Luderer, Potential and risks of hydrogen-based e-Fuels in climate change, *Nature Climate Change*, 11, 384–393, 2021.

[11]  Verband der Chemischen Industrie (VCI), Rohstoffbasis der chemischen industrie, Available: https://www.vci.de/top-themen/rohstoffbasis-chemieindustrie.jsp, 2020.

[12]  DECHEMA Gesellschaft für Chemische Technik und, carbon for power-to-X suitable sources and integration in PtX value chains, 2024.

[13]  EY Ernst & Young, bdew Bundesverband der Energie- und Wasserwirtschaft e.V., Fortschrittsmonitor 2024, 2024.

[14]  Adamovich, The 2022 plasma roadmap: Low temperature plasma science and technology, *Journal of Physics D: Applied*, 55, 373001, 2022.

[15]  Klemm, CHEMampere: Technologies for sustainable chemical production with renewable electricity and $CO_2$, $N_2$, $O_2$, and $H_2O$, *The Canadian Journal of Chemical Engineering*, 100, 2736–2761, 2022.

[16]  Fraunhofer, consent, ifeu, TU Berlin, E&R, Langfristszenarien für die transformation des energiesystems in deutschland.

[17]  Hansen, Full energy system transition towards 100% renewable energy in Germany, *Renewable and Sustainable Energy Reviews*, 102, 1–13, 2019.

[18]  Bundesministerium für Wirtschaft und Klimaschutz, Pressemitteilung strommarkt der zukunft, 2023.

[19]  DECHEMA, FutureCamp, Roadmap chemie 2050, 2019.

© 2025 World Scientific Publishing Company
https://doi.org/10.9789811282058_bmatter

# Index

## A

abundance of sodium, 176
activation energy, 215
active cell components, 42–45, 51, 53–55, 59, 62
active redox sites, 160
active sites, 164
additives, 163
aging mechanisms, 204
alloying compounds, 173
all-solid-state, 41, 46, 55, 63, 70–74, 87–88
alternative battery technologies, 140–141
alternative technology, 148
aluminum-ion batteries (AIBs), 142
ammonia, 221, 233, 235, 237
annealing, 165
anode–electrolyte interface, 212
anode materials, 171
anode overhang, 126
antimony (Sb), 173
application possibilities, 146
Archimedean spiral, 115
Arrhenius equation, 215
atomic mass, 144, 150, 152
atomic shelf, 4, 13
automotive applications, 187, 189

automotive battery systems, 187–189, 196, 198, 200
automotive requirements, 189
availability, 146
axial tabs, 120

## B

4680 battery cell, 100
battery degradation, 201
battery efficiency, 216
Battery Electric Vehicles (BEVs), 176, 221
battery lifespan, 201
battery management system, 131, 133, 192, 199, 212
battery performance, xvi
battery performance degradation, 212
battery safety, 146
battery system, 188, 190, 193–194, 196, 198
battery system integration height, 196–197
battery system level, 200
battery systems, 189, 191
bending, 129
bending of foil tabs, 99
bending of the laser-notched foil tabs, 126

256 *Index*

binder decomposition, 206
binders, 163
BIRKE-summand, 42, 63, 68, 70, 75, 78, 82
blade battery, 188, 199
building sector, 246

**C**
calendar aging, 214
capacity fade, 207
capillary wetting, 29
capturing and utilizing CO2 (CCU), 238–239
carbon coating, 159
carbon cycle, 239
catalysts, 243
cathode material delithiation, 212
cathode materials, 166, 202
cation mixing, 209
CEI instabilities, 157
Cell2Vehicle, 199
cell capacity, 95
cell concept, 187–188, 198
cell design, 36, 105, 129, 131–132, 157
cell properties, 95, 100
cell-to-module design, 190
cell-to-pack battery architectures, 130
cell voltage, 2, 9
charge-carrier ions, 145
charge density, 145, 153
charge/discharge rates, 211
charging cut-off voltage, 211
charging process, 149
chemical compounds, 223
chemical diffusion coefficient, 158
chemical energy carriers, 240
chemical industry, 221, 238, 243
chemical reduction, 165
circular economy solutions, 135
circular value creation, xv–xvii
"classical" battery, 16

compatibility, 123
composite, 173
conductivity, 36
connected energy system, 251
connection between the jelly roll and the cell housing, 121
contacting options, 120
continuous foil tab, 127, 129
control units, 189
conventional tab design, 127
conventional tabs, 129
conversion-type materials, 174
cooling system, 191–192, 199
cost, 93, 146
cost development, 17
cost-effective, 170
cost-effective battery technology, 136
costs, safety, and lifespan, 134–135
coulombic efficiency, 172
crystalline structures, 159
crystal structure, 153, 169
Current Collector Corrosion, 205
current rate, 216
cycle life, 202
cycling aging, 214
cylindrical battery cells, 96, 99, 130
cylindrical battery cells in the automotive sector, 98
cylindrical cell battery system, 188, 190, 192, 194–195, 197
cylindrical cells, 94

**D**
3D model, 31
data analysis, xvi
decarbonization, 238
defect engineering, 164
degradation mechanisms, 154, 171, 208
deintercalation, 166
dendrite formation, 157
depth of discharge, 211–212

design and format flexibility, 122
design principles, 158
desolvation energy, 153
desolvation energy barrier, 145
desolvation process, 152
diffusion, 36, 165
diffusion rates, 145, 152
diffusivity, 23
digitalization, xvi
direct electrification, 224
direct recycling, xvi
discharge energy throughput, 203
discharging process, 149
disordered structure, 172
Dissolution of Soluble Species, 205
Doping, 159
"drop-in" technology, 147

**E**
e-fuels, 221, 228, 233, 235, 238
electrical and thermal paths, 99
electrical system, 189, 192–193, 198–199
electric mobility, xv
electric vehicles, 228
electrochemical impedance spectroscopy, 102
electrochemical performance, 159
electrochemical potential, 145, 150–151
electrochemical reactions, 204
electrochemical treatment, 165
electrode balance, 214
electrode design, 24
electrode material degradation, 156
electrode pulverization, 174
electrodes, 21
electrolyte, 25
electrolyte amount, 22
electrolyte compatibility, 162
electrolyte decomposition, 156, 208, 212

electrolyte filling, 27
electrolyte oxidation, 214
electronegativity, 2
electronic conductivity, 158–159
electron transport, 164
electropositivity, 2
end-of-life batteries, xvii
energy capacity, 188
energy conversion, 237
energy density, 1, 4, 6–7, 9–13, 15, 73, 77, 93, 95–96, 98, 100–101, 107–108, 139, 150, 201, 206
energy density battery system, 187–188, 190
energy storage capacity, 222
energy system, 222–223
energy throughput, 13, 15–16
energy transition, 223
energy yield, 7–8, 10, 12
environmental impact, xvi, 141, 146
equivalent full cycles (EFCs), 203
etching, 165
exfoliation, 205

**F**
factors, 208
fast-charging applications, 199
fertilizer production, 240
folding, 129
fuel cell, 8, 16
future prospects, 17

**G**
gas formation, 156, 209
generic description, 119
generic product description, 99
geopolitical risks, 140
graphene layers, 172
Graphite, 172
Graphite exfoliation, 155
gravimetric energy density, 6, 107
green ammonia, 240

258  *Index*

greenhouse gas emissions, 223
grid stability, 227

**H**
Haber–Bosch process, 250
Hard Carbon (HC), 172
heat generation, 216
heavy-duty vehicles, 231
HF formation, 156
high operating voltage, 169–170
high specific capacities, 174
hybrid materials, 173
hydrogen, 232–234
hydrogen-based energy system, 225
hydrogen fuel cell, 12
hydrogen fuel cell vehicles, 221
hydrogen storage capacity, 225
hydropower, 245

**I**
impedance, 108
increase the volume of cylindrical
   battery cells, 98
industrial processes, 232, 242
inhomogeneities, 98–99
inhomogeneities in cylindrical battery
   cells, 98
innovative design, 99
insulating phases, 175
integrated energy system, 222
intercalation process, 153
intercalation-type anode materials,
   171
intermediate energy density, 42–43,
   45, 50, 53–55, 59–60, 63–67,
   69–72, 74–76, 78–88
internal combustion engine, 7–10, 12
internal resistance, 207
internal temperature, 104, 111
intrinsic defects, 164
ion mobility, 145, 152
ion storage, 164–165

ion transport, 164
ionic charge carrier, 145
ionic conductivity, 158
ionic radius, 144, 150, 152
Island formation, 155

**J**
jelly roll, 22
jelly roll contacting, 120
jelly roll design, 112, 119–120
jelly roll production, 124
jelly rolls with a continuous foil tab,
   130
jelly rolls with a conventional tab
   design, 130
jelly rolls with notched foil tabs, 130

**L**
lattice Boltzmann method, 27
layered structure, 166
lead-acid, 41, 44, 47, 53, 57, 64,
   84–88
lead–acid batteries, 151
LFP cells, 203
lifespan, 93
light electric vehicles (LEVs), 176
li-ion batteries, 17
lithium-air, 41, 44, 47, 49–50, 56, 64,
   78–80, 87
lithium dendrites, 212
lithium deposition, 212
lithium-ion, 41, 44–48, 51, 54–55,
   59, 63–64, 67–69, 73–74, 80–81,
   84–88, 93
lithium-ion battery, 45
lithium-metal, 41, 44, 46–47, 55, 63,
   68–69, 71, 73–75, 81, 87–88
lithium plating, 155, 205, 208, 216
lithium-sulfur, 41, 44, 47–49, 55–56,
   63, 74–78, 87
lithium–sulfur batteries (Li–S), 142
long-distance travel, 231

long-term energy storage, 225, 227
Lucas–Washburn equation, 28

## M

magnesium-ion batteries (MIBs), 142
maintenance-free operation, 16
manufacturing quality, 131, 133
manufacturing waste, xvi–xvii
market potential, 176
material choices, 157
material costs, 147
material degradation, 204
material flow, 126
material selection, 131
materials properties, 131
matrix, 14
mechanical properties, 164
mechanical stability, 194
mechanical stress, 153
mechanical structure, 190, 194, 198
mechanical system, 199
medium-term storage, 227
metal–air batteries (MABs), 8, 143
metal oxides, 174
metal–oxygen battery, 10–12
methanol, 240
mixed alloying–conversion systems, 174–175
mixed metal oxides, 167
mobility, 123
mobility sector, 221
modern batteries, 15–16
modularity, 123
morphology, 162–163
multiple electron transfer reactions, 174
multiscale model, 25
multi-tab design, 117, 119
multivalent cations, 145

## N

nanoporosity, 172
nanostructuring, 161, 168
non-active elements, 190
non-intrinsic defects, 164
notching, 128

## O

O2 phase, 167
O3 phase, 167
opening of the 4680 cell, 106
operating costs, 229
operating temperature, 190
operating voltage, 6
operational conditions, 131–132
optimal operating conditions, 210
overpotential, 13–15
oxygen battery, 7

## P

P2 phase, 167
P3 phase, 167
particle agglomeration, 161
particle cracking, 205, 209
particle fracture, 154
particle size distribution, 162
passive cell components, 42–45, 53–55, 62–63, 65, 67–77, 79–88
periodic system, 1, 5
permeability, 30, 33
phase transitions, 167
phosphorus, 174
physical exfoliation, 165
plasma processes, 243
plasma technologies, 221, 239, 241–242
policy considerations, 232
polyanion-type cathodes, 168
pore blockage, 155
pore size, 33
pore structure, 36
pore volume, 24

260  *Index*

porosity, 22, 33, 163
porous medium, 29
portable electronics, 176
pouch cell battery system, 189–191, 193, 195, 197
pouch cells, 94
power capability, 16
power density, 93, 130, 133, 152, 201
power electronic components, 193
power outputs, 15
power-to-energy (P/E) ratio, 202
power tools, 176
power-to-X (PtX), 16, 221, 223–224
pre-bending, 129
prebending of the uncoated foil tabs, 126
primary energy, 223
prismatic cells, 94
prismatic hardcase battery system, 197
prismatic hardcase cell battery system, 189–191, 193, 195
process modeling, 27
product analysis, 99
production steps, 123
production variations, 25
Prussian blue, 170
Prussian blue analog (PBA), 170
Prussian white, 171
pSEI layer formation, 156
pumped-storage power plants, 247

**Q**

quality parameters, 126
quenching, 165

**R**

raw materials, xv–xvi, 139, 144, 147
real energy density, 42–43, 54, 80, 83, 86–88
rechargeability, 4, 12–14, 16

redox-flow batteries (RFBs), 143
reducing the number of components in a battery system, 98
remaining energy densities, 187, 198, 200
removal of the jelly roll, 106
renewable electricity, 244
renewable energy sources, xv, 221–222
Replacement for Lead–Acid (PbA), 176
robust frameworks, 168
robust network, 170
rocking-chair model, 148
round-trip efficiency, 227

**S**

safe operation, 192
safety, 93, 154, 168
scalability, 123, 146
scalable cell designs, 98, 122
seasonal fluctuations, 225, 240
sector coupling, 222–223, 247, 249–250
SEI formation, 153, 208
SEI growth and dissolution, 206
SEI layer, 150–151, 154
selenium, 174
shelf, 15
short-term storage, 226
single-sided contacting, 102
single-tab and multi-tab designs, 115
single-tab design, 116, 119
size effects, 161
smart grids, 247
sodium, 144
sodium-ion, 41, 44, 47, 50–52, 56, 64, 80–81, 84, 86–87
sodium-ion batteries (SIBs), 139, 141, 148, 158

*Index* 261

sodium-ion diffusion, 166
sodium-ion electrode, 149
sodium-ion intercalation, 166
sodium ions, 142
sodium plating, 157
sodium–sulfur batteries (Na–S), 143
soft carbon, 172
solar energy, 245
solid–electrolyte interphase (SEI), 149, 157
solvent decomposition, 205
specific capacity, 4–6, 9
stability, 154
state of charge, 214
stationary energy storage systems (ESS), 176
storage technologies, 222
structural changes, 171, 204
structural disordering, 206
structural stability, 169
structure, 162
structured foil tabs, 129
structure of electrode materials, 162
sulfides, 174
sulfur, 174
supply chain disruptions, 140
surface area, 161, 163
surface coatings, 168
surface temperature, 103, 108
sustainable battery industry, 176
sustainable chemical industry, 252
synthetic fuels, 18
synthetic methane, 232
system housing, 194

**T**
4680 Tesla battery cell, 118
tab design, 115
tabless battery cell, 122
"tabless" cells, 98, 123
tabless design, 119

taping, 128
temperature, 211
Tesla design, 117
theoretical capacity, 144
theoretical energy density, 42–43, 45–49, 51–54, 63, 65–67, 69, 71–74, 76–77, 79, 81–87
thermal management, 207
thermal pathways, 103
thermal propagation, 200
thermal runaway, 190, 200, 216
thermal stability, 157
thermal system, 189–190, 192, 198–199
Tin (Sn), 173
trade-offs, 148
transfer of current and heat, 119
transition metal dissolution, 209, 214
transition metal oxide cathodes, 166
transition metal (TM) dissolution, 156
transition metals, 160
transport properties, 158–159
transportation, 246
transportation sector, 242–243
transversal tabs, 120

**U**
Universality, 123

**V**
800 V battery systems, 199
volume changes, 153, 173
volume expansion, 174–175
volumetric energy density, 6, 107

**W**
water electrolysis, 224
wavelength analysis, 105, 113

## 262  *Index*

wetting, 27
wind energy, 245
winding, 129
winding machine, 122, 126, 128, 130
winding process, 125
winding stress, 29
winding tasks, 123–124

## Z

zinc, 41, 44, 51–52, 56, 81–82, 86
zinc aqueous electrolyte, 47, 51–52, 56, 64, 81–85, 87–88
zinc-ion, 47, 51–52, 56, 64, 81–83, 87–88
zinc-ion batteries (ZIBs), 142

www.ingramcontent.com/pod-product-compliance
Lightning Source LLC
Chambersburg PA
CBHW070839150625
27790CB00017B/18